The Secret of
Everyday Things

*The Children's Classic of Scientific Learning -
Cloth, Soaps, Metals, Foods, Garden Insects and
the Physics of the World*

By Jean Henri Fabre

Translated by Florence Constable Bicknell

Published by Pantianos Classics

ISBN-13: 978-1-78987-240-8

This translation to English was first published in 1920

Contents

Introductory Note

The clearness, simplicity, and charm of the great French natural-ist's style are nowhere better illustrated than in this work, which in its variety of subject-matter and apt use of entertaining anecdote rivals "The Story-Book of Science," already a favorite with his readers. Such instances of antiquated usage or superseded methods as occur in these chapters of popular science easily win our indulgence because of the literary charm and warm human quality investing all that the author has to say.

— *Translator.*

The Secret of Everyday Things

Chapter One - Thread

UNCLE PAUL resumed his talks on things that grow and things that are made, while his nephews, Jules and Emile, and his nieces, Claire and Marie, listened to his "true stories," as they liked to call them, and from time to time asked him a question or put in some word of their own.

Continuing the subject of cotton-manufacture, he called his hearers' attention to the number of processes the raw material must go through before it emerges as finished fabric ready for making into wearing apparel, and to the countless workmen that must, from first to last, have been engaged in its production and in all the operations leading up to its final application to household uses.

"Then I should think," said Marie, "that cotton cloth would be very expensive if all those workmen are to get their pay for the time and labor they have put into its manufacture."

"On the contrary," Uncle Paul assured her, "the price is kept down to a very moderate figure; but to accomplish this surprising result two powerful factors are called into play, — wholesale manufacture and the use of machinery. The process employed for spinning cotton into the thread that you see wound on spools will help you to understand my meaning.

"You know how the housewife spins the tow that is used for making linen. First she thrusts inside her belt the distaff, made out of a reed and bearing at its forked end a bunch of tow; then with one hand she draws out the fibers and gathers them together by moistening them a little with her lips, while with the other she twirls her spindle and thus twists the loose fibers into a single strand. After she has twisted it tightly enough she winds it on the spindle, and then proceeds to draw out another length of tow from the distaff."

"Mother Annette is very skilful

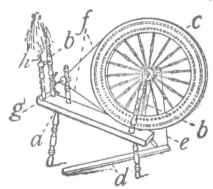

Spinning-Wheel for Flax

a, bench or stool; b, standards; c, driving band-wheel with grooved rim; d, treadle; e, rod connecting treadle with crank; f, cord-band driving the flier-spindle; g, flier; h, distaff carrying flax to be spun and, when in use, held in operator's left hand.

with the distaff," put in Claire. "I like to hear her thumb snap when she twirls the spindle. But when she spins wool she uses a spinning-wheel."

"First of all," Uncle Paul explained, "the carded wool is divided into long wisps or locks. One of these is brought into contact with a rapidly twirling hook, which catches the wool and twists it into a thread that lengthens little by little at the expense of the lock of wool, the latter being all the while held and controlled by the fingers. When the thread has attained a certain length it is wound on the spindle by a suitable movement of the wheel; and then the twisting of the lock of wool is resumed. In case of need cotton could be thus spun by hand; but, skilful as Mother Annette is at such work, cloth made from thread spun in that fashion would be enormously expensive because of the time spent in producing it. What, then, shall we do! We must resort to machinery, and in vast establishments known as cotton factories we set up hundreds of thousands of spindles and bobbins, all moving with perfect precision and so rapidly that the eye cannot follow them."

"It must be wonderful," remarked Jules, "to see all those machines spinning the cotton into thread so fast you can't keep track of them."

"Yes, those machines, surpassing in delicate dexterity the nimble fingers of the most skilful spinner, are indeed among the cleverest inventions ever produced by man; but they are so complicated that the eye gets lost among their innumerable parts. I can only point out to you the more important of these parts, without hoping to make you understand how the whole machine operates.

"First there are the cards which comb the mass of cotton into tine strips or ribbons, just as Mother Annette cards the wool she is about to spin on her wheel. These

Spinning-Wheel for Wool

a, bench; b, b', standards; c, driving band-wheel with flat rim, turned by the peg k held in the right hand of the spinner; d, cord band, crossed at e and driving the speed-pulley f; g, cord-band imparting motion to the spindle h; i, thread in process of spinning.

cards of hers, you understand, are nothing more nor less than big brushes bristling with a multitude of fine iron points. One card remains at rest and receives a thin layer of wool, after which the other is made to pass over it in such a way as to comb the wool and draw out fine locks of it, one after another. In this fashion, too, the cards in cotton factories play their part. On leaving the cards the ribbons of cotton fiber are drawn out, lightly twisted, and then wound on bobbins. Next a machine called a spinning jenny takes the partly spun cotton and twists it into thread more or less fine according to the purpose it is to serve. Finally this thread finds its way automatically to the reel, which forms it into skeins, or to the winder, which winds it into those regular balls that we can't admire too much for their perfect shape. You have doubt-less observed with what precision, what elegance, the thread is wound into a ball that the merchant delivers to you at the insignificant price of a few centimes. What human hands would have the steadiness, what fingers the skill to achieve anything comparable with this little masterpiece?"

"I know I can't begin to wind such a ball," said Marie; "it just makes a shapeless lump instead of the pretty ball I buy at the store."

"No one, depending only on his hands, could ever achieve that admirable regularity," Uncle Paul assured her. "To that end we must have machines, unvarying in their movements and working with a precision that nothing can derange.

"Thread is numbered according to its degree of fineness, the higher the

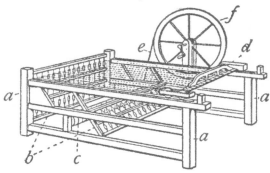

Hargreaves's Original Spinning-Jenny

a, frame; b, frames supporting spindles; c, drum driven by the band e from the band-wheel f, and carrying separate bands (not shown) which separately drive each spindle; d, fluted wooden clasp which travels on wheels on the top of the frame, and in which the rovings (the slightly twisted fibers) are arranged in due order.

number the finer the thread. Every skein and every ball being of the same length, its weight increases as the fineness diminishes. We say, then, of a particular thread that it is number 200 when it takes two hundred skeins or balls to make half a kilogram in weight, and that it is number 150 when it takes one hundred and fifty to make up the same weight."

Chapter Two - Pins

"AFTER thread, come the needle and its companion the pin. I shall take up the latter first, because its manufacture will help us to understand that of the needle, which is rather more complicated.

"The things most often used by us are not seldom those of whose origin we are ignorant. What is there more convenient, more often used, than the needle and the pin? What could take their place if we were deprived of them? We should be reduced to Claire's makeshift that day we went on a picnic and she tore a hole in her apron and fastened the edges together with a thorn from the hedge. We might also, as do those savage tribes that have no manu-factured articles, shred an animal sinew or a strip of bark into fine thongs to serve as thread and sew with a sharp-pointed bone for a needle. We might replace the pin by a fish bone."

"That would be a funny sort of gown," exclaimed Marie, "sewed with thongs of bark or the sinews of an ox; nor should I care much to have my hair fastened with codfish bones."

"Yet there are even to-day savage tribes that have nothing else; and often the great ladies of ancient times had nothing better: they used rude pins made of metal or little splinters of bone. Advance in the manufacturing arts has given us the pin, with its pretty round head, at a price so moderate as to be almost negligible, the needle with its fine point and its admirable suitability to our use, and thread of remarkable strength and fineness. Now let us learn how pins are made.

"Pins are made of brass, which is composed of copper and zinc. Copper is the red metal you are familiar with in copper kettles, zinc the grayish-white metal of watering-pots and bath-tubs. Mixed together they form brass, which is yellow.

"The first step is to reduce the copper to wire the size of a pin. This is done by means of a draw-plate, a steel plaque pierced with a series of holes, each smaller than the preceding. A little brass rod is thrust into the largest hole and forcibly drawn through it. In passing through this hole, which is a little too small for it, the metal rod becomes correspondingly thinner and longer. It is then thrust into a still smaller hole and again drawn out, becoming once more thinner and longer in the process. This operation is continued, passing from one hole of the draw-plate to the next smaller, until the wire acquires the desired fineness.

"While we are on the subject note this fact — that all metal wires, whether of iron, copper, gold, or silver, are made in the same way: namely, by being passed through the draw-plate.

"The brass wires are now put into the hands of the cutter, who gathers several of them into a bundle and then, with a strong pair of shears, cuts them all into pieces twice the length of a pin.

"These pieces must next be sharpened at both ends by means of a steel grindstone which has its grinding-surface furrowed like a file, and which turns with the prodigious velocity of twenty-seven leagues an hour. The man charged with this work, whom we will call the sharpener, sits on the ground in front of his grindstone, legs crossed in tailor-fashion. He takes in his fingers from twenty to forty pieces, spreads them out regularly in the shape of a fan, and brings all these branching tip-ends simultaneously into contact with the grindstone, at the same time twirling them in his fingers so that the tip is worn off equally all around and the point made even. The reverse tips are sharpened in the same way.

"But this first process merely produces points in the rough, so to speak; the sharpener retouches and finishes them on a finer grindstone. Finally the pieces sharpened at both ends are arranged several together and cut in two in the middle with one clip of a pair of shears. Each half, known as a shank, now lacks only a head in order to become a complete pin.

"This heading process is the most difficult part of the whole operation. On a slender metal shaft, very smooth and slightly larger than the pins, a thread of brass is tightly wound in a spiral, after which the shaft is removed, leaving a long corkscrew with its turns touching one another. A cutter of consummate skill in this delicate work, which demands at the same time so much precision and so much swiftness, divides this corkscrew into small pieces, each containing just two turns. Each of these pieces is a head.

"The workman who is to put them in place and fasten them takes the shanks one by one and plunges them haphazard, pointed end first, into a wooden bowl full of heads. The shank is drawn out with a head strung on it, which the operator pushes with his fingers to the unpointed end. He immediately places it on a little anvil having a tiny cavity into which the head fits; then by means of a pedal moved by the operator's foot a hammer provided with a similar cavity comes down, strikes five or six little blows, and behold the head firmly fixed.

"As a finishing touch the pins have still to be coated with tin. To this end they are boiled with a certain proportion of this metal in a liquid capable of dissolving it and depositing it in a thin layer on the brass. After being thus coated they are washed, dried on cloths, and finally shaken up with bran in a leather bag in order to heighten their polish.

"It only remains to stick the pins in paper in regular rows. A kind of comb with long steel teeth pierces the paper with two lines of holes. Workwomen known as pin-stickers are charged with the delicate task of inserting the pins one by one in these holes. A skilled pin-sticker can insert from forty to fifty thousand pins a day.

"Including some details that I omit, the manufacture of a pin requires fourteen different operations, and consequently the cooperation of fourteen workmen, all of consummate skill in their part of the operation. Nevertheless the manufacture is so rapid that these fourteen workmen can make twelve thousand pins for the modest sum of four francs." [1]

[1] Since the foregoing was written automatic machinery has been invented which greatly facilitates the manufacture of pins. Pointing, heading, and papering are now done with great rapidity by such machinery, and hand-work is almost entirely dispensed with. — *Translator.*

Chapter Three - Needles

TAKE from a case one of the finest needles, examine its sharp point, its tiny, almost imperceptible eye, and note finally the polish, the shine. Tell me if this pretty little tool, so perfect in its minuteness, would not seem to require for its manufacture the superhuman fingers of a fairy rather than man's heavy hands. Nevertheless it is robust workmen with knotty fingers blackened by the forge and covered with great calluses that do this most delicate work. And how many workers does it take to make one needle! — one only! For the manufacture of a pin, I have already told you, it takes fourteen different workmen; for the manufacture of a needle it requires the cooperation of one hundred and twenty, each of whom has his special work. And yet the average price of a needle is about one centime. [1]

"The metal of needles is steel, which is obtained by adding carbon to iron heated to a very high temperature. Under this treatment iron changes its nature a little, incorporating a very small quantity of carbon and thus becoming exceedingly hard, but at the same time brittle. A needle must be very hard in order not to bend under the pressure of the thimble forcing it through the thickness of the material on which the seamstress is at work, and also in order that the point may not become blunted, but always retain the same power of penetration. Steel, the hardest of all the metals, is the only one that fulfils these conditions of resistance; neither copper nor iron nor the precious metals, gold and silver, could replace it. A gold needle, for example, in spite of its intrinsic value, would be useless, becoming blunted and twisted before using up its first needleful of thread. Steel alone is suited to the manufacture of needles, though unfortunately this metal is brittle, and the more so the harder it is."

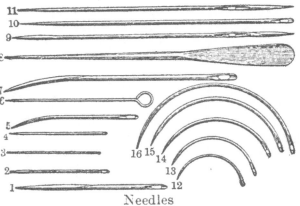

Needles

Various shapes and sizes used by sailmakers and upholsterers.

"But I should think," Marie interposed, "What since steel is so hard it ought not to break." "You will think otherwise if you listen to me a while. Hardness is the degree of resistance that a body opposes to being cut, scratched, worn away by another. Of two bodies rubbing against each other the harder is that which cuts the other, the softer is that which is cut. Steel, which scratches iron, is harder than iron; in its turn glass is harder than steel, because it can cut the steel without being cut by it. But a diamond is still harder than glass, since it scratches glass and glass cannot scratch it. In fact, a diamond is the hardest of all known substances: it scratches all bodies and is scratched by none. Glaziers take advantage of this extreme hardness: they cut their panes of glass with the point of a diamond."

"I have heard," said Claire, "What a diamond placed on an anvil and struck with a hammer stands the blows without breaking and penetrates into the iron of the anvil, it is so hard."

"That is a great mistake," replied Uncle Paul. "A diamond breaks like glass, and he would be very ill-advised who should submit the precious stone to the proof of the hammer. At the first blow there would be nothing left but a little worthless dust. You see by these different examples that hardness and brittleness are often united. Steel is very hard, glass still harder, the diamond the hardest of all substances; nevertheless all three are brittle. That explains to you why needles of excellent steel, which gives them their rigidity and power of penetration, nevertheless break like glass in clumsy fingers.

"Now I come to the subject of manufacture, from which the properties of steel turned us for a moment. The metal is drawn out into wire by means of a draw-plate; then this wire, several strands at a time, is cut into pieces twice the length of a needle, just as in pin-making. The pieces are pointed at each end, first on a revolving sandstone similar to an ordinary grindstone, then on a wooden wheel covered with a thin layer of oil and a very fine, hard powder called emery. Imagine glass reduced to an impalpable powder and you will have a sufficiently correct idea of what emery is. The first process gives a more or less coarse point; the second sharpens this point with extreme nicety.

"The pieces thus pointed at both ends are cut into two equal parts, each one of which is to be a needle. The workman then takes in his fingers four or five of these unfinished needles, spreads them out like a fan and puts the large end of them on a little anvil; then with a light blow of the hammer he slightly flattens the head of each. It is in this flattened end that later on the eyelet or hole of the needle will be pierced."

"But you just told us, Uncle," Marie interrupted, "that good steel is brittle, the same as glass; yet the workman flattens the head of his needles with a hammer without breaking anything."

"Your remark is very timely, for before going further we have to take note of one of the most curious properties of steel. I must tell you that it is only by tempering that this metal becomes hard and at the same time brittle. Tempering steel is heating it red-hot and then cooling it quickly by plunging it into cold water. Until it undergoes this operation steel is no harder than iron; but, to compensate for this softness, it can then be hammered, forged, and in fact worked in all sorts of ways without risk of breaking. Once tempered, it is very hard and at the same time so brittle that it can never henceforth stand the blow of a hammer. Accordingly needles are not tempered until near the end of the process of manufacture; before that they are neither hard nor brittle and can be worked as easily as iron itself.

"If you look at a needle attentively you will see that the head is not only flattened but also hollowed out a little on each side in the form of a gutter or groove which serves to hold the thread. To obtain this double groove, the workman places the needles, one by one, between two tiny steel teeth which, moved by machinery, open and shut like two almost invisible jaws. Bitten hard by the shutting of these two teeth, the head of the needle is indented with a groove on each side.

13

"Now the eye must be pierced, an operation of unequalled delicacy. Two workmen cooperate in this, each equipped with a steel awl whose fineness corresponds with the hole to be made. The first places the head of the needle on a leaden block, puts the point of his instrument in the groove on one side, and, striking a blow with the hammer on the head of the awl, thus obtains not a complete hole but merely a dimple. The needle is then turned over and receives a similar dimple on the other side. The other workman takes the needles and with the aid of his awl removes the tiny bit of steel that separates the two dimples. Behold the eye completely finished.

"Probably no work requires such sureness of hand and precision of sight as the piercing of the eye of a needle. Certainly he has no trembling fingers or dimmed eyesight who can, without faltering, apply his steel point to the fine head of a needle, strike with perfect accuracy the blow of the hammer, and open the imperceptible orifice that my eyes can scarcely find when I want to thread a needle."

"There are needles so small," remarked Marie, "that I really don't see how any one can manage to make an eye in them."

"This incomprehensible achievement is mostly the work of astonishingly skilful children. So skilful, indeed, are some of them that they can make a hole in a hair and pass a second hair through this hole."

"Then the needle's eye," said Emile, "which seems such a difficult piece of work to us, is only child's play to them."

"Child's play indeed, so quick and dexterous are they at it. And they have still another kind of dexterity that would astonish you no less. To make the needles easier to handle in the process of manufacture, they must be placed so that they all point the same way; but as in passing from one operation to another, from one workman to another, they become more or less disarranged, it is necessary to arrange them in order again, all the points at one end, all the heads at the other. For us there would be no way but to pick them up one by one; with these children this delicate task is but the work of an instant. They take a handful of needles all in disorder, shake them in the hollow of the hand, and that is enough; order is reestablished, the heads are together, the points together.

"The eye completed, the next process is tempering, to give the steel its required hardness. The needles are arranged on a plate of sheet-iron, which is then placed on red-hot coals. When sufficiently heated, the needles are dropped quickly into a bucket of cold water. This produces in them the hardness characteristic of steel, and its accompanying brittleness.

"As a finishing touch the needles must be polished till they shine brightly. In parcels of fifteen or twenty thousand each they are sprinkled with oil and emery and wrapped up in coarse canvas tied at both ends. These round packages, these rolls, are placed side by side on a large table and covered with a weighted tray. Workmen or machinery then make the tray pass back and forth over the table unceasingly for a couple of days. By this process the packages, drawn this way and that by the tray, roll along the table, and the needles, rubbing against one another, are polished by the emery with which they are sprinkled.

"On coming out of the polishing machine the needles, soiled with refuse of oil and detached particles of steel, are cleaned by washing with hot water and soap. It only remains now to dry them well, discard those that the rude operation of polishing has broken, and finally wrap with paper, in packages of a hundred, those that have no defect. The most celebrated needles come from England, but needles are also made in France, at Aigle in the department of Orne." [2]

[1] Nearly one fifth of a cent in our money. — Translator.
[2] Since Fabre wrote, the manufacture of needles, like that of pins, has undergone important changes and improvements through the application of machinery. — *Translator.*

Chapter Four - Silk

THE culture of the silkworm having been explained by Uncle Paul in one of his previous talks, [1] he now confined himself chiefly to the structure of the cocoon and the unwinding of the delicate silk thread composing it.

"The cocoon of the silkworm," he began, "is composed of two envelops: an outer one of very coarse gauze, and an inner one of very fine fabric. This latter is the cocoon properly so called, and from it alone is obtained the silk thread so highly valued in manufacture and commerce, whereas the other, owing to its irregular structure, cannot be unwound and furnishes only an inferior grade of silk suitable for carding.

"The outer envelope is fastened by some of its threads to the little twigs amid which the worm has taken its position, and forms merely a sort of scaffolding or openwork hammock wherein the worm seeks seclusion and establishes itself for the serious and delicate task of spinning its inner envelop. When, accordingly, the hammock is ready the worm fixes its hind feet in the threads and proceeds to raise and bend its body, carrying its head from one side to the other and emitting from its spinneret as it does so a tiny thread which, by its sticky quality, immediately adheres to the points touched. Without change of position the caterpillar thus lays one thickness of its web over that portion of the enclosure which it faces. Then it turns to another part and carpets that in the same manner. After the entire enclosure has thus been lined, other layers are added, to the number of five or six or even more. In fact, the process goes on until the store of silk-making material is exhausted and the thickness of the wall is sufficient for the security of the future chrysalis.

"From the way the caterpillar works you will see that the thread of silk is not wound in circles, as it is in a ball of cotton, but is arranged in a series of zigzags, back and forth, and to right and left. Yet in spite of these abrupt changes in direction and notwithstanding the length of the thread — from three hundred to five hundred meters — there is never any break in its continuity. The silkworm gives it forth uninterruptedly without suspending for a moment the work of its spinneret until the cocoon is finished. This cocoon has an average weight of a decigram and a half, and it would take only fifteen or twenty kilograms of the silk thread to extend ten thousand leagues, or once around the earth.

"Examined under the microscope, the thread is seen to be an exceedingly fine tube, flattened and with an irregular surface, and composed of three distinct concentric layers, of which the innermost one is pure silk. Over this is laid a varnish that resists the action of warm water, but dissolves in a weak alkaline solution. Finally, on the outside there is a gummy coating which serves to bind the zigzag courses firmly together and thus to make of them a substantial envelop.

"As soon as the caterpillars have completed their task, the cocoons are gathered from the sprigs of heather. A few of these cocoons, selected from those that show the best condition, are set aside and left for the completion of the metamorphosis. The resulting butterflies furnish the eggs or 'seeds' whence, next year, will come the new litter of worms. The rest of the cocoons are immediately subjected to the action of very hot steam, which kills the chrysalis in each just when the tender flesh is beginning slowly to take form. Without this precaution the butterfly would break through the cocoon, which, no longer capable of being unwound because of its broken strands, would lose all its value.

"The cocoons are unwound in workrooms fitted up for the purpose. First the cocoons are put into a pan of boiling water to dissolve the gum which holds together the several courses of thread. An operator equipped with a small broom of heather twigs stirs the cocoons in the water in order to find and seize the end of the thread, which is then attached to a reel in motion. Under the tension thus exerted by the machine, the thread of silk unwinds while the cocoon jumps up and down in the warm water like a ball of worsted when you pull at the loose end of the yarn. In the heart of the unwound cocoon there remains the chrysalis, inert, killed by the steam.

"Since a single strand would not be strong enough for the purpose of weaving, it is usual to unwind all at once a number of cocoons, from three to fifteen and even more, according to the thickness of the fabric for which the silk is destined; and these united strands are used later as one thread in the weaving machines.

"As it comes from the pan the raw silk of the cocoon is found to have shed its coating of gum, which has become dissolved in the hot water; but it is still coated with its natural varnish, which gives it its firmness, its elasticity, its color, often of a golden yellow. In this state it is called raw silk and has a yellow or a white appearance according to the color of the cocoons from which it came. In order to take on the dye that is to enhance its brilliance and add to its value, the silk must first be cleansed of its varnish by a gentle washing in a solution of lye and soap in warm water. This process causes it to lose about a quarter of its weight and to become of a beautiful white, whatever may have been its original color. After this purifying process it is called washed silk or finished silk. Finally, if perfect whiteness is desired, the silk is exposed to the action of sulphur, as I will explain to you when we come to the subject of wool.

"Cocoons that have been punctured by the butterfly, together with all scraps and remnants that cannot be disentangled and straightened out, are carded and thus reduced to a sort of fluff known as floss-silk, which is spun on the distaff or the spinning-wheel very much as wool is treated; but even with the utmost pains the thread thus obtained never has the beautiful regularity and the soft fineness of that

which is furnished by unwinding the cocoon. It is used for fabrics of inferior quality, for stockings, shoe-laces, and corset-laces.

"The silkworm and the tree that feeds it, the mulberry, are indigenous to China, where silk-weaving has been practised for some four or five thousand years. To-day, when the highly prized caterpillar is dying out in our part of the world, China and its neighbor Japan are called upon to furnish healthy silkworm eggs. Silk-culture was introduced into Europe from Asia in the year 555 by two monks who came to Constantinople with mulberry plants and silkworm eggs concealed in a hollow cane; for it was strictly forbidden to disseminate abroad an industry that yielded such immense riches."

[1] See "The Story-Book of Science."

Chapter Five - Wool

"WE live," continued Uncle Paul, "on the life of our domestic animals. The ox gives us his strength, his flesh, his hide; the cow gives us her milk besides. The horse, the ass, the mule work for us; and when death overtakes them they leave us their skin for leather with which to make our footwear. The hen gives us her eggs, and the dog places his intelligence at our disposal. But if there is one animal that, more than another, comes to us from the good God above, it is surely the sheep, the gentle creature that yields us its fleece for our garments, its skin for our warm coats, its flesh and its milk for our nourishment. But its most precious gift is its wool.

"From wool are made mattresses, and it is also woven into cloth such as merino, flannel, serge, cashmere, and, in short, all the various fabrics best fitted for protecting us from the cold. It is by far the most desirable material for wearing apparel, cotton, notwithstanding its importance, coming only second, and silk, valuable though it is, being very inferior in respect to serviceability. More than with anything else we clothe ourselves with what we strip from the innocent sheep; our finery comes for the most part from its fleece."

"But wool is very far from beautiful on the creature's back." commented Claire; "it is all matted and dirty, often fairly covered with filth."

"It must take a good many processes," remarked Marie, "to change that foul and tangled fleece into the beautiful skeins of all colors with which we embroider such pretty flowers on canvas."

"Yes, indeed, very many," rejoined Uncle Paul. "I have already told [1] you how sheep are washed and sheared, and how the washing leaves the fleece white or brown or black according to the color given to it by nature. White wool can be dyed in all possible tints and shades, from the lightest to the darkest, whereas brown or black wool can take only somber hues. White wool, therefore, is always preferable to any other; but, beautiful as it is when freshly washed and relieved of all impurity, it is still far from having that snowy whiteness so desirable if it is to remain undyed. It is bleached by a very curious process which I will now describe to you.

"You have all doubtless observed that when sulphur burns, with a blue-violet flame, it gives forth a pungent odor that irritates the mucous membrane of the nose and throat and causes a fit of coughing."

"That must be what we smell when we light a match," Claire interposed. "If you breathe in the least little whiff of it, it is perfectly horrid."

"Often enough it has set me to coughing unless I was on my guard," remarked Emile.

"Yes, that is it," their uncle replied. "Sulphur, in burning, becomes an invisible substance which is dissipated in the atmosphere and betrays its presence only by a detestable odor of the most pungent quality. Invisible, impalpable, like the air itself, this something that we know merely as a disagreeable smell constitutes nevertheless a real substance the existence of which cannot be doubted by any one who has once been thrown into a fit of coughing by inhaling it. It is called sulphurous oxide, a new name to you and one to be kept in mind. It will be worth your while to remember it, as you will presently see."

"Sulphurous oxide, then," said Marie, "is burnt sulphur; and it is something that can be neither seen nor felt, but that nevertheless does really exist. Whoever breathes it is immediately convinced of its existence by the penetrating odor and by the fit of coughing that follows."

"To what possible use," continued Uncle Paul, "can we turn this disagreeable gas, this invisible substance that makes you cough worse than if you had the whooping-cough! I will tell you. Despite its repulsive qualities, it is what we have to depend upon for giving to wool the whiteness of snow. An example will demonstrate its efficacy to you. Go down to the meadow and pick me a bunch of violets."

The violets were soon gathered from under the hedge bordering the meadow. Then Uncle Paul put a little sulphur on a brick, set it afire, and held the bunch of violets, which he had slightly sprinkled with water, over the fumes. In a few moments the flowers, attacked by the sulphurous gas ascending from the blue flame, lost their color and turned perfectly white. The change from violet to white was plainly visible to the eye.

"How curious that is!" exclaimed Jules. "Must see how the violets whiten as soon as they come over the flame and feel the sulphurous oxide, as you call it. Some were half white and half blue; but the blue has disappeared and now the bunch is all white, without having lost any of its freshness to speak of."

"Let us now, "suggested Uncle Paul, "try one of the red roses there on the mantelpiece."

Accordingly the rose was held over the burning sulphur, and its red color faded away just as the blue of the violets had faded, giving place to white, much to the wonder of the children, who watched with breathless interest this marvelous transformation.

"That will suffice for the present," Uncle Paul resumed. "What I have just shown you with violets and roses might be demonstrated with innumerable other flowers, especially red and blue ones: all would turn white on being exposed to the sulphur fumes. You will

Head of Merino Ram,
Before and after shearing.

understand, then, that these fumes, which we call sulphurous oxide, have the peculiar property of being able to destroy certain colors and hence to act as a bleaching agent.

"If, therefore, you wish to bleach wool, to remove the slight natural discoloration that stains its whiteness, you proceed exactly as you have just seen me do with the violets and roses. In a room with all its doors and windows carefully closed the wool in its natural condition — that is, before it has been spun into yarn — is hung up and a good handful or two of sulphur is set on fire in an earthen bowl. The room then becomes filled with sulphurous oxide and the wool turns a beautiful white."

"Would wool that is naturally brown or black turn in that room full of sulphur smoke!" asked Marie.

"No," was the reply; "its color is too fast to yield to the action of sulphurous oxide. Only white wool is subject to this action, under which it becomes immaculate. By the same process the straw of which hats are made is bleached, also skins used for gloves, and silk.

"Wool varies in value according to the different kinds of sheep that have produced it, some being coarse, some fine and silky, some made of long hairs, and some of short. The most highly esteemed, that which is used in weaving

19

fine fabrics, comes from a breed of sheep raised chiefly in Spain and known as merino sheep. Finally, a goat native to the mountainous countries of central Asia, the goat of Cashmere, furnishes a downy fleece of extreme fineness, an incomparable wool from which the most costly stuffs are manufactured. This goat wears, under a thick fur of long hair, an abundant down which shields it from the rigors of winter and is shed every spring. At that season the animal is combed and the down is thus detached separate from the rest of the hairy coat."

[1] See "Our Humble Helpers."

Chapter Six - Flax and Hemp

"THE inner coating of the stalk of flax and hemp, as I have already told you, is composed of long filaments, very fine, flexible, and strong, which are used like cotton in the manufacture of various fabrics. Flax gives us such fine

fabrics as cambric, tulle, gauze, and laces of various kinds; hemp furnishes us stronger stuffs, up to the coarse canvas used for making sacks. Flax, as you have already learned, is a slender plant with small flowers of a delicate blue. It is sown and reaped annually, and is raised especially in northern France, in Belgium, and in Holland. The first of plants to be used by man for making fabrics, it was turned to account by the people of Egypt, the land of Moses and the Pharaohs, for the furnishing of linen bands with which to wrap the mummies that have been reposing in their sepulchres more than four thousand years. So carefully, indeed, were they embalmed and then wrapped in linen and enclosed in chests of aromatic wood that to-day, after the lapse of centuries upon cen-

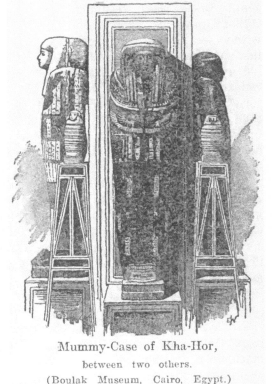

Mummy-Case of Kha-Hor,
between two others.
(Boulak Museum, Cairo, Egypt.)

turies, the contemporaries of the ancient kings of Egypt, of the Pharaohs in other words, are found intact, though dried up and blackened by time."

20

"But in spite of all these precautions," objected Claire, "surely the mummies must have gone to decay if they were buried in the ground!"

"For that reason," replied her uncle, "they were not buried; they were laid away in orderly rows in spacious halls hollowed out of the solid rock of mountains. These mortuary halls, to which dampness never penetrated and the air had but little access, have kept for us intact, swathed in their linen bands, the bodies of the ancient Egyptians."

Uncle Paul next took up the subject of hemp, relating the history of its cultivation in Europe from early times and describing its appearance, with its small green flowers and its slender stalk about two meters in height. He explained that, like flax, it is grown both for its fibrous stem and for its seed, known as hempseed, which is used as a favorite food for certain singing-birds. From the seed are obtained hempseed oil and hempseed cake, the latter being sometimes fed to cattle.

"And what is flaxseed good for?" asked Emile.

"From the seeds of flax," answered his uncle, "is obtained by pressure an oil called linseed oil, which can be used for lighting, but is chiefly employed in painting. For culinary purposes it is almost worthless, being of no use at all unless very fresh, and even then of but moderate value. Its principal use, as I said, is in painting, because of its quality of slowly drying and thus forming a sort of varnish which holds fast the pigment with which it is mixed. The coat of paint that overlies, for example, the woodwork of doors and windows is made of linseed oil in which has been stirred a mineral powder, white, green, or any other color chosen by the painter. When flaxseed is ground it yields a pow-

Flax

Blossoming plant and section of seed-vessel.

der much used for poultices, being of an unctuous nature soothing to pains.

"When hemp and flax are ripe they are harvested and the seeds are detached either by threshing or by passing the seed-bearing ends of the stalks through a strong iron-toothed comb. The comb is set up across the middle of a bench on which two workmen seat themselves astride, one at each end, facing the comb. Then, by turns, they draw each his handful of flax or hemp through the comb, thus separating the seeds from the stalks.

"Next comes the operation known as retting, whereby the fibers of the bark are rendered separable from the rest of the stem and from one another. The gummy substance holding them together has to be disintegrated either by prolonged exposure in the field, where the flax or hemp is turned over from time to time, or, more expeditiously, by soaking the stalks in Waaler, after first tying them into bundles. The resulting putrefaction liberates the fibers. Drying, breaking, and hackling them complete the separation of the fibers from the useless substance of the stem and their reduction to a condition in which they are ready for use.

"I will add that the fibrous part of hemp, as you may know already, is far coarser than that of flax. The filaments of the latter are so fine, that one gramme of tow, spun on the wheel, makes a thread nearly one hundred and fifty meters long. Nevertheless, this product of man's skill, this linen thread that seems to reach the limit of fineness, is very coarse indeed when compared with what is furnished by the caterpillar and the spider. The highest degree of delicacy attainable by our fingers with the aid of the most ingenious machinery is but an enormous cable in contrast with the thread manufactured by a despised little worm. A single gramme of the silkworm's thread, as we find it in the cocoon, represents a length of two thousand meters, whereas the finest of linen thread of the same weight represents only one hundred and fifty.

Hemp

Male (1) and female (2) plants of hemp. *a*, male flower; *b*, female flower; *c*, embryo.

"But even the slender filament spun from the silkworm's spinneret is incomparably coarser than the spider's thread, the achievement of that master artisan the very sight of whom evokes from you senseless outcries of alarm. To weave the airy textures intended to catch their prey, such as flies and gnats and similar small game, as also to line the dainty little sachets that hold their eggs, spiders on their part produce a sort of silk. The silk matter is contained in liquid form in the spider's body and is forced out as required through four or five little nipples called spinnerets, situated at the end of the insect's stomach, each of these nipples being perforated with many tiny holes, the total number of which for a single spider is reckoned at about a thousand. Hence the spider's thread as it leaves the insect's body is not a single strand, but a cord of a thousand strands, although we commonly consider it of almost infinitesimal minuteness. Our finest sewing silk is a stout cable in comparison, and a human hair has the thickness of ten twisted spider's threads or, in other words, of ten thousand combined -elementary filaments of spider-silk. How inconceivably fine then must be a thread that needs to be multiplied ten thousand times in order to equal a human hair in size! The larger spiders that live in woods weave webs of remarkable amplitude, requiring each at least ten meters of thread, or ten thousand meters of the elementary filament emitted by a single aperture of the spinneret. But to make the entire web the spider uses up only a tiny drop of liquid silk, of which it would take hundreds of similar drops to weigh a gramme. What machine of human invention or what fingers could spin for us a thread of any such inconceivable fineness!"

[1] See "The Story-Book of Science."

Chapter Seven - Weaving

"EXAMINE a piece of cloth, woolen, cotton, or linen, and you will see that it is composed of two sets of threads which cross one another, each thread passing alternately over and under a transverse one. Of these two sets one is called the warp, the other the woof or weft, and their crossing produces the woven fabric, or cloth.

"The work of weaving these threads into cloth is done by means of a loom. I will try to describe to you an old-fashioned hand-loom, which is much simpler in construction than the modem power-loom. A solid wooden framework supports a cylinder in front and one at the back, and these cylinders are turned each by a crank whenever needed. The front cylinder, its crank within reach of the operator seated ready for

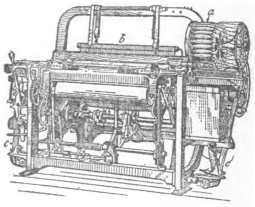

Modern Power-Loom

work, receives the woven stuff a little bit at a time; the other, fixed at the opposite end of the machine, is wound with threads in regular order side by side. These threads will form the warp of the cloth, and they are stretched with careful regularity between the two cylinders the whole length of the machine. They are divided into two sets, the odd-numbered threads forming one set, the even-numbered threads the other. Two heddles hold the two sets and keep them separate without possibility of intermingling. A heddle is a series of very line metal wires, or it may be simply threads, stretched vertically between two horizontal bars."

"The heddles are those two gridiron things in the middle of the loom?" asked Claire.

"Precisely. At every wire or thread of the heddle there is passed, in order, through an eye or ring, one of the strands composing the two sets of the warp.

Now notice that by means of two pedals or levers placed under the operator's feet the two heddles can be made to rise and fall alternately. In this alternate movement they draw by turns, up and down, one the even threads and the other the odd threads of the warp.

"While the warp is thus slightly open, all the even threads on one side, all the odd on the other, the operator sends the shuttle through the space separating the two sets. The shuttle is a piece of boxwood, well polished so as to slide easily, tapering at each end, enlarged in the middle, and provided with a cavity that holds a bobbin of thread fixed on a very mobile axle. This thread unwinds automatically with the throwing of the shuttle, and is left lying between the two sets of threads of the warp. Then with a pressure on one of the pedals the order of these sets is reversed, the threads that were above passing below, those below coming uppermost, and the shuttle sent in the opposite direction leaves another thread stretched across. This thread furnished by the shuttle and passing by turns from right to left and from left to right between the two lines of the warp forms what is called the woof or weft of the cloth."

Shuttle

a, body of shuttle; *b*, yarn wound on bobbin *d*; *c*, eye through which the yarn is led and then passed out through hole *f*; *e, e*, metal points.

"So the feet, "said Marie, "by pressing the pedals make the odd and even threads of the warp move up and down, while the hands, sending the shuttle from right to left and then from left to right, interlace the thread of the woof with the warp."

"That is the double movement the operator has to learn — the pressing of each foot in turn on the pedals and the sending of the shuttle from one hand to the other. But in order that the cloth may acquire sufficient firmness, with no open spaces between the threads, these two movements are supplemented by a third. A comb-like instrument called a reed is used to "beat up' or press close together the threads of the woof after every two or three passages of the shuttle through the warp, or sometimes after every passage, according to the nature of the fabric.

"Such, in short, my dear children, is the process by which all our woven fabrics of two sets of intercrossed threads are made, cloth, linen, taffeta, calico, and a great many others."

Chapter Eight - Woolen Cloth

I HAVE just given you a general description of the art of weaving. Now I propose to add some details relating to the more important products of the loom. And first let us take up woolen cloth.

"Woolen cloth is woven of woolen yarn. As it comes from the spinning-wheel or spinning-jenny this yam has certain surface irregularities, little bristling fibers standing up and crinkling with the natural curliness characteristic of wool. In this state the yarn would check the easy gliding of the shuttle, which must shoot back and forth with great rapidity; and thus the work would be rendered laborious and the woven fabric wanting in evenness of texture. The surface must be made as smooth and uniform as possible, the fluff flattened and held down the whole length of the thread. This is done by means of a preparation or facing with which the threads of both the woof and the warp are coated. In this preparation are glue, which holds down the fluff, and oil, which makes the surface slippery.

"Thus it is that, as it comes from the loom, cloth is badly soiled, carrying as it does a coating of glue and ill-smelling oil. Before these impurities become seats of decay the cloth must be cleaned, and it must be done as soon as possible. The operation is carried out in a fulling-mill, which consists of a series of heavy wooden clubs or beaters set in motion by means of a wheel turning in a stream. The beaters alternately rise and then fall with all their weight to the bottom of a trough continually sprinkled by a jet of clear water. The cloth is placed in the trough where, the clubs beat it one after another for whole days. But this energetic beating is not enough; the glue would disappear, but not the oil, which is more tenacious and on which water has no effect. Accordingly, recourse is had to a sort of rich earth, fine and white, which has the property of absorbing oil. It is called fullers' earth."

"That rich earth could be used then for taking out grease spots?" queried Marie.

"It is used for that purpose. All you have to do is to cover the grease spot for a while with a layer of fullers' earth made into paste, and the grease will disappear, being absorbed by the clay. In many countries it is used instead of soap for washing clothes."

"What a funny kind of earth!" Claire exclaimed. "I should like to wash with it. What is it like?"

"It is a white clay, greasy to the touch, taking a polish when smoothed with the finger-nail, and mixing readily with water, to which it gives a soapy look. In France the best-known fullers' earth is found in the departments of Indre, Isère, and Aveyron.

"Beaten with this earth for a number of hours by the heavy clubs of the fulling-mill, the cloth loses the oil with which it is impregnated. Soap-suds and finally pure water finish the cleaning.

"But the part performed by the fulling-mill is not limited to cleaning the cloth; it also shrinks the goods to half the original width and nearly half the length. In this connection I will call your attention to a precaution familiar to every good housewife. Before cutting out a garment she is careful to wet the cloth so as to shrink it as much as possible. If this precaution were not taken, the garment would shrink so in the first washing that you couldn't get into it."

"That is what happened to Emile's linen trousers," said Jules. "They came out of the wash so short they hardly reached to his knees."

"A rope shrinks, too, when it is wet," remarked Marie. "Once, after a rain, the clothes-line in our back yard shrank so that it pulled out the hooks it was fastened to."

"That reminds me of a little anecdote," said Uncle Paul. "When shortened by being wet, a rope exerts so strong a pull that not only can it extract hooks, but it can even lift immense weights. It is said that Pope Sixtus the Fifth, when he was about to erect in one of the public squares of Rome an obelisk brought from Egypt at great expense, ordered under pain of death the most profound silence during the operation, so anxious were the operating engineers on account of the enormous weight to be moved. I will tell you, before going further, that obelisks are tall, slender, four-sided columns engraved with a multitude of figures and crowned by a small pyramid. They are in one piece, of a very hard and fine-grained stone called granite. Their height, not counting the pedestal that supports them, may reach fifty meters, and their weight may range between ten thousand and fifteen thousand hundredweight. Judge, then, whether the erection of this ponderous mass upon its pedestal did not present difficulties.

"To operate in perfect unison the numerous ropes, pulleys, and levers used for raising the immense piece, absolute silence was necessary so that not a word should distract the workmen's attention. The square was crowded with curious idlers watching this mighty exertion of mechanical power. Complete silence reigned, every one bearing in mind the pope's order. But when the raising of the obelisk had proceeded half-way, the enormous stone refused to go further and remained leaning with all its weight on the ropes. Everything was at a standstill. The engineers, at the end of their resources, saw their gigantic task

Fullers' Teazel
a, scale of the receptacle;
b, corolla.

threatened with failure, when suddenly from the midst of the crowd a man's voice rose at the peril of his life. 'Wet the ropes!' he cried. 'Wet the ropes!' They wet the ropes and the obelisk soon stood upright on its pedestal. The tension of the cordage when soaked with water had of itself done what an army of workmen had failed to accomplish."

"And what happened to the man who broke the silence?" asked Emile.

"The pope willingly pardoned him, you may be sure. But let us return to our subject of woolen cloth. You can now easily understand what happens when this cloth is wet. It is made of crossed threads, each one of which, on being soaked with water, acts like a rope, that is to say it becomes shortened. The result of this is a closer texture. On drying, the cloth does not return to its original state, as a rope when dry resumes its former length; it remains close, because the threads, held in position by their interlacing, are not free to slip. Thus by being put through the fulling-mill, where it is beaten and wet at the same time, the cloth which was at first loose enough to show the day-

26

light between its meshes, becomes a firm piece of goods with warp and woof close together.

"The two sides of a piece of cloth are not the same: one, called the wrong side, shows the crossed threads of the fabric, otherwise known as the thread; the other, called the right side, is covered with a fine, even nap, all lying the same way. This nap is obtained by means of a kind of rude brush made of the thorny burs furnished by a plant called teazel, or fullers' teazel.

"Teazel lives from one to two years. Its stalk, which attains the height of a man, is armed with strong hooked thorns and bears, at a certain distance apart, pairs of large leaves, each pair forming a cup more or less deep in which rain gathers. Growing from the main stem are six or seven branches, each terminated by a strong elongated head or bur composed of hard scales sharply pointed and recurved at the end in the shape of a fine hook. The plant is cultivated expressly for its burs, which are used in great quantities in cloth-manufacture. It would be difficult to replace this natural brush with any similar tool made by our hands, for nothing could give the same degree of needed stiffness and suppleness combined. Five or six of these burs are placed side by side so as to form a brush, which is drawn over the cloth always in the same direction. The thousand hooks of the teazel, each as fine as the slenderest needle, but elastic and supple, seize the tiny fibers of surface wool lying between the threads, and pull them out, laying them one on the other, all pointing the same way. The result of this operation is the nap which on the right side of a piece of cloth covers and hides the thread.

"But this nap is still imperfect: its tiny fibers are of unequal length, some long, some short, at haphazard, just as the hooks of the teazel brush drew them from the threads. To make it all smooth and even, it must be shorn; that is to say, large broad-bladed shears are used to pare down the surface of the cloth so as to leave the nap all of the desired length. This completes the essential part of the work. Sedan, Louviers, and Ellbeuf are the chief cloth-manufacturing towns of France."

Chapter Nine - Moths

"IN our houses," continued Uncle Paul, "we have a redoubtable enemy to woolen cloth and everything else that is made of wool — an enemy that in a very short time will reduce a costly garment to rags and tatters unless we are on our guard against the ravager. Therefore it is worth our while to make the acquaintance of this devourer of woolen goods, this despair of the housewife, in order that we may hunt it down with some success. You know the little white butterflies that come in the evening, attracted by the light, and singe their wings in the lamp-flame. They are the ravagers of woolen fabrics, the destroyers of broadcloth and other woolen stuffs."

27

"But those little butterflies, "objected Claire, "are feeble little creatures to tear in pieces anything so substantial as broadcloth."

"And for that very reason it is not the butterfly itself that we are afraid of; the delicate little flutterer is perfectly harmless. But before turning into a butterfly it is first a caterpillar, much like the silkworm; and this caterpillar is endowed with a voracious appetite that makes it gnaw substances apparently uneatable, such as wool, furs, skins, feathers, hair. To the caterpillar and its butterfly we give the name of moth."

"There are caterpillars, then, that eat cloth and even hair?" asked Marie.

"There are only too many of them," was the reply. "One of these caterpillars, one that some day will turn into a pretty little butterfly all powdered with silver dust, would feast right royally on your woolen frock; and another would find much to its taste your fur tippet, which keeps your shoulders warm in winter."

"There can't be much taste in a mouthful of fur, I should think, and it must be pretty hard to digest."

"I don't deny it, but those caterpillars have stomachs made expressly for that sort of diet, and they accommodate themselves to it very well. A worm that eats fur and digests hair knows nothing in the world so good, and one that gnaws old leather would turn away with aversion from a juicy pear, a piece of cheese, or a slice of ham, all of them repugnant to its taste. Every species has its preferences and, according to its mode of life, possesses a stomach designed to find nutriment in substances apparently far from nutritious. On the moth's bill of fare are skins, leather, wool, woolen

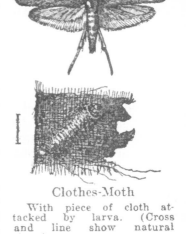

Clothes-Moth

With piece of cloth attacked by larva. (Cross and line show natural sizes.)

cloth, fur, and hair. The larva does not merely feed on these materials, but it also makes from them a movable house, a sheath that covers its body, leaving the head free, and this house it carries about with it.

"All butterflies of the moth class have narrow wings bordered with an elegant fringe of silky hair and folded lengthwise on the back in repose. Of the three principal species the distinguishing characteristics are as follows:

"The woolen moth has black upper wings tipped with white, while the head and lower wings are white. Its grub, or larva, is found in woolen goods, and it is there that it makes for itself a sheath from the bits of the gnawed fabric.

"The fur moth has silver-gray upper wings with two little black dots on each. Its grub lives in fur goods, which it denudes, a hair at a time.

"Finally, the hair moth lives, in its grub state, in the curled hair used for stuffing cushions and couches. In color it is of a uniform pale red.

"The moth most to be feared is the one that feeds on woolen cloth. Let us discuss its habits more in detail, for in spite of its ravages you will admire, with me, the skill it displays in making itself a coat. To protect itself so that it may live in peace, the grub fashions for itself a sheath from the bits of wool cut and chopped with its sharp little teeth. In thus cutting down these up-standing hairs, one by one, the worm shears the cloth and makes a thread-bare spot. The shearman himself could not have operated with such nice precision. But there is nothing so disfiguring in new cloth as these shorn spots showing here and there the warp and woof of the fabric, while all the rest retains its velvet finish. Furthermore, the mischief is not always confined to the shorn spots: too often it happens that the tiny destroyer attacks the threads themselves and makes holes here and there in the cloth, so that the latter is found to be nothing but a worthless bundle of rags. The bits of wool thus cut away serve the worm either as food or as building-material for its movable house, its sheath.

"This latter is most deftly put together, consisting on the outside of tiny bits of wool fastened together with a little liquid silk emitted by the worm, and on the inside of silk alone, so that a tine lining protects the creature's delicate skin from all rough contact."

"Just think of it," exclaimed Jules; "the detestable devourer of our woolen clothes lines its own coat with silk!"

"And that is not all," continued Uncle Paul. "The little creature indulges in the luxury of assorted colors. Its coat takes the hue of the cloth in process of destruction, and thus there are white coats, black coats, blue coats, and red coats, according to the color of the material. If this latter happens to be of variegated tints, the worm takes a bit of wool here and a bit there, making for itself a sort of harlequin outfit in which all the colors represented are min-gled at haphazard.

"Meanwhile the worm continues to grow and its sheath becomes too short and too tight. To lengthen it is an easy matter: all that is required is to add new bits of wool at the end. But how is it to be made larger?"

"If I had to do it," Claire replied, "I should run my scissors down length-wise, and in the opening I should insert another piece."

"The ingenious insect seems to have taken counsel of Claire, or of an even better tailor," said Uncle Paul. "With its teeth for scissors it cuts open its coat all down its length and inserts a new piece. So skillfully is this insertion made, so neatly are the seams sewed with the silk, that the most expert of dressmakers would find it hard to pick any flaw in the workmanship."

"These moth-worms must be very skilful, I admit," said Marie, "but I shouldn't like to have them practice their art on my clothes. How are they to to be prevented?"

"To protect garments from moths it is customary to place in our wardrobes certain strongly scented substances such as pepper, camphor, tobacco. But the surest safeguard is to inspect the garments frequently, shaking them and beating them and exposing them to the sun. All moths love repose and darkness. Garments that are shaken occasionally and hung in the light are not at all to their taste; but those that are laid away for months or years in a dark place offer gust the kind of snug retreat they are looking for, the ideal abode for the raising of a family. Go to your chests of drawers and your wardrobes very often and shake, air, and brush the contents; then you will have no moths. Vigilance is here worth more than pepper and camphor. Finally, kill all the little white butterflies you see fluttering about your rooms."

"But those little butterflies do no harm whatever, you told us, "objected Emile. "It is only the worms that gnaw our clothes."

"True enough; but those butterflies will lay eggs by the hundred, and from every egg will come a devouring worm. The destruction of the flying moth means therefore deliverance from some hundreds of future moths."

Chapter Ten - Napery

"HEMP is woven into coarse material for towels and sacks, and even into finer material for sheets, chemises, table-cloths, and napkins. From flax is obtained still finer goods for the same purposes. Sometimes the same material contains both hemp and flax. Thus the goods known as cretonnes, manufactured at Lisieux and its' environs, have the warp of hemp and the woof of flax. Sometimes, again, it is cotton that is mixed with hemp. Ticking, for example, is a very close fabric used for making coverings for bolsters and also men's summer clothes. Generally it is all hemp, but certain grades have a cotton woof. So it is that the three kinds of vegetable fiber — hemp, flax and cotton — can be used two together in the same material, which gives goods of greater variety and better adapted to the infinite uses for which they are destined.

"Goods of this sort generally bear the name of the country that produces them: such are the goods called Brittany, Laval, Valenciennes, Saint-Quentin, Voiron. Others are named after their inventor, as Cretonne, which derives its name from a manufacturer, Creton, who centuries ago gained a great reputation for linen-manufacture. One kind of linen, very fine and close, used for handkerchiefs and various articles of attire, such as veils, collars, and cuffs, is called batiste in honor of Baptiste Chambrai who was the first to make this material, and who introduced its use about five centuries ago.

"Material composed only of hemp and flax, either separate or together, is commonly called linen. Certain qualities distinguish these goods from cotton. To a delicate skin they have a cool and soothing feeling, whereas cotton, owing to its nap, which is slightly rough, produces a kind of tickling that may be positively disagreeable. Thus a cotton handkerchief irritates nostrils that have been made sensitive by a prolonged cold; but a linen handkerchief has not the same objectionable quality. And again, for dressing wounds it is customary to use linen or hemp bandages and lint obtained from old rags of the same material, since cotton, no matter how fine, and soft, would only increase the irritation of the wound by its rough contact with the quivering flesh. Finally, hemp and linen as used for underwear keep the skin in a state of coolness that is very agreeable in the heat of summer, but which may under certain conditions prove very disagreeable. Let perspiration be checked, let the body, poorly protected by its cool covering of hemp or linen, cool off quickly, and we are in serious danger. Cotton, on the contrary, stimulates the skin slightly, keeps it warm, and affords better protection when perspiration is arrested. In this respect it is preferable to linen and hemp. But I will come back to this subject after some details that I wish to give you in a subsequent talk on the conservation of heat.

"As soon as hemp has been spun into thread by the long and patient labor of the distaff, it is sent to the weaver, who coats it with a preparation of glue to facilitate the play of the shuttle, stretches it on his loom in parallel lines, and weaves it as I have already explained to you, each foot pressing in turn one of the pedals that operate the warp, and the two hands throwing, one to the other, the shuttle which stretches the thread of the woof between the two sets of warp threads. A good washing cleans the cloth, removing the preparation I have referred to and all impurities contracted during the weaving. But that is not enough to produce the beautiful white cloth that the housewife cuts into shirts and sheets. Hemp and flax are, in fact, naturally of a light reddish tint, so firmly fixed that only after repeated washings will it entirely disappear; which explains why sheeting is whiter as well as softer the longer it is used.

"As a first step in bleaching, the linen is spread on the ground in a well-mown field, where for whole weeks it remains exposed to the daylight and to the damp night air. The prolonged action of air and sun, dryness and dampness, at length fades the reddish color, which subsequent washings will, little by little, finally remove altogether.

"This bleaching by exposure to sun and rain is very slow. Moreover, when the operation has to be carried on uninterruptedly and on a large scale it is very costly, because it renders unproductive considerable stretches of land. Consequently in hemp, linen, and especially cotton factories recourse is had to means that are at once more energetic and more expeditious. You remember how easily and economically wool and silk are bleached by burning a few handfuls of sulphur, thus generating a gas called sulphurous oxide. It is only necessary to expose wet wool and silk for a few hours to the action of this gas to give them the dazzling whiteness of snow."

"Is that the way hemp, flax, and cotton are treated?" asked Marie.

"Not quite, although the method employed much resembles that used for wool. Sulphurous oxide would have no effect here, so difficult is it to destroy the natural color of hemp, flax, and cotton. Something stronger, something more drastic, must be used."

"But that sulphur smoke is pretty strong; it pricks your nose like needles, and makes you cough till the tears come."

31

"Yet it is nothing in comparison with the drug used for bleaching. This drug is also a gas — that is to say, a substance as impalpable as air, but at the same time a visible gas, for it has a light greenish color. It is called chlorine. If you breathe a whiff of it, you are immediately seized with a violent cough such as you would never get in winter, however cold it might be. The throat contracts painfully, the chest is oppressed, and you would die in frightful torture if you inhaled this formidable gas three or four times in succession. You can see, then, what precautions one must take in factories not to expose oneself to the terrible effects of chlorine."

"And what does it come from, this gas that strangles people if they breathe ever so little of it?" asked Claire.

"It comes from common salt, the same salt with which we season our food. But I must add that in salt it is not found all by itself; it is mixed with another substance which renders it harmless, even wholesome. Once freed from this partnership it is murderous, a frightfully destructive agent. I am sorry I cannot show you its astonishing power in destroying colors; but nothing prevents my telling you about it. Imagine a sheet of paper not only covered with characters traced by the pen but daubed all over with ink. Now plunge this into chlorine gas, and writing and ink-blots all disappear instantly, leaving the sheet of paper as white as if it had never been used. Suppose, again, you put chlorine into a bottle of ink. The black liquid fades quickly and soon there is nothing left but clear water.

"After this you can understand that the material to be bleached has to be subjected to the action of chlorine for only a few moments in order to turn whiter than through long exposure in the field."

"If the deep black of ink is destroyed so quickly," remarked Marie, "the pale reddish tinge of hemp or linen is not likely to hold out very long."

"Wool and silk," Claire observed, "ought to be bleached that way too: it would be much quicker."

"The manufacturers are very careful not to follow any such method," was the reply. "This gas corrodes wool and silk, soon reducing them to a mere pulp."

"And yet cotton, flax, and hemp can stand it," Claire rejoined.

"Yes, but their resistance to the action of drugs has not its equal in the world, and this resistance gives them a very peculiar value. Think in how many ways cloth of this sort is used, and what severe treatment it undergoes: repeated washing with corrosive ashes, rubbing with harsh soap, heating, exposure to sun, air, and rain. What then are these substances that withstand the asperities of washing, soap, sun, and air, that even remain intact when all around them goes to decay, that brave the drugs used in manufacturing and emerge from these manifold tests softer and whiter than before? These almost indestructible substances are hemp, flax, and cotton; and they have no rivals."

Chapter Eleven - Calico

"IT now remains for me to tell you about the principal weaves of cotton. First there is percale, which has a firm and close texture and a smooth surface, and is much used for shirts, curtains, covers, and sometimes for table and bed linen. Ornamented with colored designs, it is also used for dresses.

"Percale, diminutive of percale, as its name shows, is a fabric of inferior quality and of transparent texture, being very loosely woven. Its thread is flat and its surface fluffy and plush-like, whereas in percale the thread is round and the surface smooth. It lacks firmness and does not last long. It is used chiefly for lining.

"Common calico is less fine, less firm, and cheaper than percale, but is used in general for the same purposes.

"Muslin is a very fine, soft, light material, the most delicate of all cotton goods. There are some muslins that for fineness almost rival the spider's web, and of which a piece several yards long could be contained in an eggshell. Among muslins are classed nainsook, organdie, and Scotch batiste.

"Cotton has a decided superiority over flax and hemp in that it readily takes any desired color or ornamental design, the dyes being quickly absorbed, lastingly retained, and shown off to the best advantage. Who does not know these admirable goods in which the most varied and brilliant colors are artistically combined and our garden flowers are reproduced in astonishing perfection? Some of these prints are decorated with flowers such as no garden could furnish. Cotton alone lends itself to this richness of coloring, hemp and flax being absolutely inadequate. Cotton goods ornamented with colored designs are called prints, and they first came from India, where their manufacture has been known for a very long time. To-day the factories of Rouen, Mulhouse, and England supply these goods to the whole world. It will interest you to hear about some of the methods employed by the calico-printer in his delicate work. How were those beautiful designs obtained, so clear and bright, that ornament the most inexpensive dress? That is what I propose to tell you in a few words.

"First the fabric is bleached with the greatest care so that no dinginess of its own shall dim the brightness of the colors to be received. Energetic washing, over and over again, and the powerful bleaching agent I have just spoken of, namely chlorine, make the cotton perfectly white.

"Now comes an operation that would fill you with astonishment if you happened to see it: it is the operation of singeing. I must tell you to begin with that all cotton thread, however perfect the spinning-machine producing it, is covered with a short down or fluff consisting of the tip-ends of the vegetable fibers standing up by their own elasticity. At the time of weaving this fluff is laid flat by means of a preparation of glue, so as to leave perfect freedom for the play of the shuttle; but now this preparation, which would greatly interfere with the setting of the colors, has disappeared to the last trace, and the fluff of the threads stands up free again. Well, on such goods, all bristling with tiny filaments, the colored designs would not take well; there would be unevenness of tint, ill-defined outlines; cracks and seams, in fact. The surface must be as clean and smooth as a sheet of paper. It would be well-nigh impossible to obtain with shears such as are used for shearing woolen cloth a surface as smooth as the subsequent operations demand. Accordingly it is the custom to have recourse to the use of fire. The material is passed with the necessary rapidity before a broad jet of flame, which thoroughly burns off every bit of fluff without in the least damaging the fabric itself. Nothing is more extraordinary to the novice than to see a piece of calico or percale or even muslin passing through the menacing curtain of flame without catching fire."

"And who would not be surprised!" exclaimed Marie. "I should think the delicate fabric would certainly catch fire."

"You would think so, but there is no danger if the material passes quickly along and does not give the heat time enough to penetrate beyond the fluff. Let ns dwell a moment on this peculiarity, which will enlighten us concerning a very remarkable attribute of cotton.

"You know what happens when you put the end of a piece of cotton thread into the lamp flame. The part thrust into the flame is consumed at once, but the fire spreads no farther and goes out just at the point where the thread ceases to be enveloped by the flame. With a piece of linen or hemp thread the result would be a little different: the thread would continue to burn more or less beyond this limit. That is on account of the different manner in which cotton on the one hand and flax and hemp on the other act under the influence of heat. Cotton is rather impervious to heat; flax and hemp, on the contrary, offer only feeble resistance to its spreading. I will not say any more on this point now, but will return to it some day with the necessary details.

"The few facts I have given you are sufficient to explain what takes place in the singular operation of singeing. If the fabric passes through the fire quickly and at an even rate of speed, the flame envelops it on both sides and even traverses the meshes, burning off all the fluff without injuring the threads themselves, because the heat has not time to spread further.

"To banish once for all your incredulity in regard to this singeing process I will show you an experiment which, indeed, has no close connection with the calico-printer's art, but which illustrates the different degrees of inflammability possessed by different substances. What should you say if I were to tell you that live coals can be placed on the finest muslin without burning it in the least?"

"I should say, 'Seeing is believing,'" answered Claire.

"Then you shall see, Miss Incredulous. Take a piece of muslin, as fine as you please, and wrap it tightly around one of the brass balls that ornament the top of the stove. Tie it securely underneath with a string so that the muslin will touch the metal in every part. Now take from the open fire a live coal and apply it with the tongs to the muslin that covers this kind of doll's head."

Claire followed these directions with scrupulous care: the live coal was touched to the muslin, and, greatly to the surprise of all the children, the delicate fabric remained perfectly intact.

"Go still farther," commanded Uncle Paul. "Take the bellows and make the live coal burn as brightly as you like, letting it rest on the muslin while you do so."

Claire worked the bellows, the coal became red-hot all over, and still the muslin underwent no change, appearing to be quite incombustible.

"Why, this is unbelievable!" she cried. "How is it that the muslin can stand the touch of a live coal without burning the least bit?"

"What protects it from the fire," replied her uncle, "is the metal underneath. The brass, a substance easily penetrated by heat, takes to itself the heat of the coal and leaves none for the muslin, which is much harder to heat. But if the delicate fabric were by itself, it would burn at the first touch of the live coal."

Several times during the day Claire repeated this experiment by herself, each time more astonished than before at this strange incombustibility.

Chapter Twelve - Dyeing and Feinting

"ARRIVED at this point, the cotton cloth is ready to receive the colors. This operation involves the use of means so varied and technical knowledge so above your understanding that I should not be understood if I undertook to enter into some of the more elaborate details."

"I supposed, on the contrary," said Claire, "that it was a very simple thing and that the colors were put on the cloth with a paint-brush just as I should do, though not very well, on a sheet of paper."

"Undeceive yourself, my dear child: the paintbrush has nothing to do with printed cottons, nor yet with the other kinds of goods embellished with colors. Done with a paint-brush, the designs would not be lasting, but would disappear with the first washing. The slightest rain would make the colors run and would turn them into horrible shapeless blots. To resist water, and sometimes vigorous washing with soap, the colors must penetrate the fabric thoroughly and become embodied in it.

Gall-fly (female), natural size; b, male antenna magnified

"Let us see how this result is arrived at, taking black for an example. This color is obtained in various ways, notably with ink, the same that we use for writing. Well, if we immersed a strip of white material in this liquid it would come out black, but the color would have no staying power. A little rinsing in water would remove most of the ink, and the small amount remaining would

35

give only a pale tint, very insufficient and soon washed out. In order to give a deep and lasting black the ink, when it is applied to the fabric, must be in an unfinished state, and it must finish itself in the cloth; the ingredients of which it is composed must mingle and become ink in the very substance of the threads. Under these conditions the black, made on the spot and permeating the minutest fibers of the cotton, acquires all the desired fixity and intensity. Let us, then, before going further, examine the ingredients of which ink is composed.

"There are found growing on oak-trees certain globular formations of about the size of a billiard ball and with the appearance of fruit. But they are not really fruit; they have nothing in common with acorns, the real fruit of the oak. They are excrescences caused by the sting of a tiny insect known as the gall-fly. This insect stings the leaf or tender twig with a fine gimlet that arms the tip-end of its stomach, and in the microscopic incision it deposits an egg. Around this egg the sap of the tree gathers, finally forming a little ball which by degrees becomes as hard as wood. The insect hatched from the egg develops and grows in the very heart of this ball, whose substance serves it as food. When it has grown strong enough it pierces the wall of its abode with a small round hole through which it escapes. That is why you see most of these balls pierced with a hole when they fall to the ground toward the end of autumn. These round excrescences are called gall-nuts or oak-apples, and they furnish one of the ingredients of ink, one of the materials used for dyeing black.

"The other ingredient is green copperas. At the druggist's you may have seen a substance looking something like broken glass of a light green color with spots of rust. That is green copperas. It is obtained by dissolving iron in an excessively corrosive liquid known as oil of vitriol or sulphuric acid. This terrible liquid, so dangerous in inexperienced hands, dissolves iron as easily as water dissolves salt or sugar; and in this solution, after a certain time, crystals are formed. It is this crystallization that gives us our green copperas, a substance having none of the dangerous qualities of the oil of vitriol used in making it, but one that can be handled without the least risk, though its taste is most disagreeably tart. This substance dissolves very readily in water.

"That is all that is needed for making ink. Let us boil a handful of pounded gallnuts in water; we shall obtain a pale yellowish liquid. Also, let us dissolve some copperas in water, and the latter will turn a very pale green with a yellowish tinge. What will happen on mixing the two liquids, one yellow and the other green? Nothing very remarkable, it would seem. And yet no sooner do these two liquids mingle than there is produced a deep black, the very color of ink."

"And so the ink comes all of a sudden." said Marie.

"Yes, the very instant the gallnuts and the copperas meet in the water that holds them in solution. If ink for writing is desired, however, a little different method must be followed so as not to have an excess of water, which would weaken the color. To the liquid resulting from boiling the gallnuts there would simply be added the undissolved copperas with a little gum to give brilliance to the ink. I have gone out of my way in describing to you this mingling of two liquids in order to show you the more clearly how the union of two substances, each having little or no color, can produce a color totally unlike either of the original shades. From two liquids, one a pale yellow, the other a pale green, you have just seen ink spring into being with startling suddenness. Remember this phenomenon, for it will explain certain results of dyeing that are very astonishing to the uninitiated. Bear in mind that sometimes perfectly colorless substances can produce magnificent colors by being united.

"Now that we know how ink is made, let us return to the subject of dyeing black. We put a piece of percale to soak in water in which gallnuts have been boiled; then, when it is well saturated, we take it out and dry it. What, now, will be its color!"

"It will be the color of the gallnut water," answered Claire; "that is to say, a pale and dirty yellow."

"Right; but if the fabric thus saturated is dipped into a solution of copperas, what will happen?"

"That is not hard to guess, "was the reply. "The copperas, finding gallnut dye on the surface and in the texture of the cloth, all through it in fact, will immediately form ink, which will color the percale black."

"And more than that," added Marie, "The dye will penetrate the goods evenly in every part, since the saturation with gallnut water extends to the very tiniest thread of the material."

"Yes; and so you see that in this way the black dye is formed on the spot, in the very heart of the cotton threads. Thus we obtain a fast color, all the necessary conditions being complied with.

"A great many other colors — red, violet, brown, yellow, lilac, no matter what — are produced in similar fashion. First the material to be dyed is saturated with a solution that will develop the desired color, or make it spring into existence, on encountering another solution, and will set the color by making it one with the fabric itself. This preparatory substance which in a second operation is to mix with the dyestuff so as to develop and fix the color is called the mordant and varies in kind according to the tint desired, so that by changing the mordant different colors may be obtained with one and the same coloring matter."

"In the way you have just explained to us," said Marie, "we get cloth dyed all one color. I should like to know how patterns of several colors on a white ground are made."

"That is done by printing or stamping. Imagine a small wooden block or board on which is engraved in relief the design to be reproduced. Clever engravers skilled in all the details of ornamental design prepare these blocks, which are sometimes veritable masterpieces of art; and it is these that constitute the calico-printer's all-important equipment.

"To take a simple example, let us suppose the workman proposes to put a black design on a white ground. On a large table in front of him he has the piece of percale which unrolls as he needs it; in his right hand he holds the printing-block. The engraved design, which stands out in relief, he moistens slightly with a fine solution of gallnuts, and then applies the block to the goods. The parts thus touched are the only ones impregnated with this preparation, the rest of the percale remaining as it was before. He continues thus, each time dipping the engraved face of the block into the gallnut preparation, until the piece of cloth has received the impression throughout its entire length.

"That done, all that is necessary is to dip the goods into a solution of copperas to make the design appear in black, since the ink forms wherever the wooden mold has left a deposit of gallnut water, all other parts remaining white."

"It is simpler than I had thought," said Claire, "and much simpler than using a paint-brush, as I supposed at first must be the way."

"The operation can be made still simpler. As a rule the coloring matter and the mordant, that is to say the substance that brings out the color and fixes it, act only under the influence of heat. Accordingly, the process is as follows. The two ingredients, mordant and coloring matter, are mixed together and reduced to a fine pap with which the engraved surface of the block is moistened and then immediately applied to the fabric. The preparation thus deposited gives only one color, one alone, determined by the nature of the mordant and the coloring matter. If the design is to be multicolored, as many blocks must be used as there are tints, each block representing only the part of the design having the color it is to imprint on the goods. Thus the piece of percale passes through the workman's hands once for red, again for black, a third time for violet, in fact as many times as there are colors in the design, however little they may differ from one another."

"It must be a very delicate piece of work," remarked Marie, "to put the different parts of the design exactly in the right place so as to get from them all a pattern "with the various colors joining perfectly and never overlapping."

"The calico-printer's skill makes light of this difficulty. The design comes out as clear as a painter could make it with his brushes. To complete the description in a few words, when all the colors have been applied, the fabric is removed to a closed room where it receives a steam bath. Heat and moisture aiding, each dye mixes with its mordant, which incorporates it with the fabric, and beautiful bright tints spring forth as by enchantment where the engraved blocks had left only a sorry-looking daub."

Chapter Thirteen - Dyestuffs

"COULDN'T you tell us, Uncle Paul," said Marie, "how all those colors we see on printed goods are obtained? There are such beautiful reds, blues, violets, that real flowers can hardly compare with them."

"Yes, I will tell you. Let us first take madder, the most valuable of dyestuffs on account of the beauty and fastness of its colors. It is the root of a plant cultivated in France, chiefly in the department of Vaucluse, and of about the size of a large feather, reddish yellow in color. The preparation it undergoes before being used in dyeing consists simply in reducing it to very fine powder and purifying it as much as possible.

"Madder by itself imparts absolutely no color to any stuff, be it silk, wool, cotton, or whatever other you please. One might boil for days at a time a piece of percale in water containing powdered madder; the fabric would remain white. For the color to take form and impress itself on the stuff there is some essential condition lacking."

"No doubt it needs what you call the mordant, that substance that mixes with the dye to bring out the color and fix it on the goods, just as gallnuts bring out the black of ink by mixing with copperas."

"That is it: it lacks the mordant. In the case of madder this is sometimes iron-rust and sometimes a white substance resembling starch and called alumina, which is obtained from very pure clays. If the material to be dyed is first saturated with a strong solution of alumina, it takes a dark red tint on being dipped into boiling water containing a proper amount of powdered madder. If the quantity of alumina is small,

Madder

1. branches with flowers and fruits; 2, the rhizome; a, blossom; b, the pistil; c, two different fruits.

the resulting tint is simply rose-color. Thus by varying the proportion of the mordant any shade can be given to the fabric, from the deepest red to the palest pink.

"With iron-rust for mordant other colors quite unlike the preceding are developed. A liberal quantity of rust gives black, a small quantity violet — always with madder, be it understood. Finally, if the mordant is a mixture of alumina and rust the color produced is a chestnut brown, intermediate between red and black, and varying in shade according to the proportions of the ingredients used. You see, then — and it cannot fail to surprise you — that with a single dyestuff, madder, it is easy to obtain a numerous series of

39

tints ranging from dark red to pale pink, from deep black to delicate violet, and including also the chestnuts or mixtures of red and black.

"Let us suppose the calico-printer has stamped the different mordants on the goods with his printing-block and has artistically grouped them to obtain bouquets of flowers. This done, the cloth appears merely soiled with unsightly blotches, the iron-rust showing with its dirty yellow, while the alumina, being colorless, remains invisible. But as soon as the piece is plunged into a boiling bath of madder each mordant attracts to itself the coloring matter dissolved in the water, incorporates it, and with it forms such and such a color according to its nature. The reds, pinks, blacks, violets, chestnuts, all come out at the same time before our astonished eyes, which at first might imagine themselves beholding the birth of enchanted bouquets."

"If you had not explained this curious operation," said Claire, "I should have been astonished to see those magnificent printed bouquets taking shape all by themselves in the confusion of that boiling vat."

"Then it is in that one vat," added Jules, "containing only water and madder, that there are formed all at the same time the reds, pinks, and violets for the flowers, the chestnuts for the bark of the branches, and the blacks for the shadows. The bouquets lack only the green of the leaves to be complete."

"Madder does not give green; another substance and another operation are necessary to obtain that color. Nevertheless, who could fail to perceive the importance of madder, that one substance which furnishes so many hues, all remarkable not merely for their beauty but also for their unequaled permanence? No other dyestuff contains in itself so many excellent qualities."

Woadwaxen or Dyers'
Greenweed

"And the other colors — blue, for example — how are they obtained?" asked Marie.

"The most lasting blue is the product of a plant called the indigo-plant. It is too cold in our part of the world to raise this plant, but it grows on the warm, damp plains of India. It is the leaves that are used. They are green at first, but if they are allowed to go to decay in water containing a little lime, a substance having a superb blue color and called indigo is formed from them. "A very beautiful yellow remarkable for its fastness is prepared from a plant that grows around here and is known as woadwaxen or dyers' greenweed, bearing flowers closely resembling those of the mignonette, so famous for its sweet odor. By mixing this yellow with blue we obtain the green that Jules spoke of as needed for the leaves of the bouquets.

40

"A small, rather ugly-shaped insect gives the dyer his most beautiful reds. It is known as the cochineal, and it lives all its life in one spot, as do the lice of our rose bushes. It infests a fleshy plant whose branches are flattened in the shape of a palette and studded with tufts of thorns. This plant is known under the names of nopal, cactus, Indian fig, and Barbary fig. Mexico produces most of the cochineal. The insect is gathered from the nopal, killed by immersion in boiling water, and dried in the sun. It then looks like a little wrinkled seed. About one hundred and forty thousand insects are required to make Female cochineal one kilogram in weight.

Female Cochineal

Dried specimen of commerce. (Line shows natural size.)

It is only necessary to boil the cochineal in water to obtain a red liquid which deposits as sediment the beautiful coloring matter known by the name of carmine. Wool and silk are dyed scarlet with cochineal.

"I will conclude with a few words on the brightest, clearest of all dyestuffs, but unfortunately, also, the most changeable, the most evanescent. Recall the splendid hues now given to wool, silk, and above all to ribbons. The rainbow alone can rival them. Now do you know the origin of these colors, so pure, so bright, so charming to the eye? They come from a horrid, malodorous substance called coal-tar.

"In the first place you should know that illuminating gas is obtained by heating coal red-hot, in large iron vessels to which no air is admitted. The heat liberates at the same time gas for lighting and tar which is set aside by itself; there is then left a kind of coal, light, shiny, full of holes, and called coke. Let us turn our attention to the tar only, which despite its disgusting appearance is one of the most marvelous products known to the manufacturing world. By treating it first in one way, then in another, and after that in still another, there are obtained from it a number of very different substances, some resplendent like mother-of-pearl or the scales of fishes, others white and powdery like fine flour, and still others resembling limpid oil and having in certain instances a strong and disagreeable odor and in others an aromatic fragrance. When this separation is complete, we have at our disposal various substances which further processes will transform into colors of all kinds. One of these substances derived from coal-tar, and at first a colorless oil, becomes an azure blue that would not disfigure the wing of the most gorgeous butterfly; another, at first a floury powder, reproduces exactly the colors yielded by madder; a third gives shades of red beside which the queen of flowers, the rose, would look pale. But one capital fault [1] is common to most of these splendid colors obtained by man's skill from the somber coal: hardly any of them can stand the least washing without injury, and even light alone fades them quickly.

"Colors that are really fast, those that last as long as the fabric that bears them, and can without fading stand light and soap, are particularly the colors obtained from madder, the browns and blacks into the making of which gall-

41

nuts have entered, the blues from indigo, and the yellows of woadwaxen. Beware of a dye that charms the eye but turns dim under the first ray of sunlight or with the first washing."

[1] This fault has now been corrected. — *Translator.*

Chapter Fourteen - Heat-Conduction

"To give you a thorough understanding of the part played by woven fabrics in our clothing and coverings I must now call your attention to certain attributes of heat. The little I can tell you on this subject will be illustrated in many cases by important practical applications. So be very attentive.

"A firebrand can with impunity be taken up in the fingers by one end if the fire is confined to the other; there is no risk of our being burnt, even though the fingers be very close to the ignited part."

"I have often noticed that," said Jules, "when poking a corner of the fire in winter, and when to save time I have used my fingers instead of the tongs."

"But one could not, without getting burnt, grasp the apparently cold end of a piece of iron, even a pretty long one, if red at the other end. Neither could one take hold of the handle of a flat-iron warming on the heater. The hand must be protected by a thick holder.

"These two familiar examples, charcoal and iron, show us that heat is not conducted with the same readiness by all substances. It easily makes its way through iron, which becomes very warm a long distance from the part directly heated; but it is with difficulty that it permeates charcoal, which remains cold a short distance from the ignited section.

"In this respect all substances are divided into two classes: those that are easily permeated by heat or that conduct it well, and those that are difficult for heat to permeate or that conduct it poorly. The first are called good conductors, such as iron; the second, poor conductors, such as charcoal.

"Among the good conductors are all the metals — iron, copper, silver, gold, and so on. Non-metallic substances, such as wood, charcoal, brick, glass, and the various kinds of stone, are, on the contrary, poor conductors. Conductivity is still poorer in powdery substances, such as ashes, earth, sawdust, and snow; also in fibrous substances, such as cotton, wool, silk, and hence the fabrics woven from these materials."

"Then," said Marie, "the holder with which we grasp the handle of a flat-iron keeps the hand from being burnt because it prevents the heat from going farther. It is a poor heat-conductor."

"Yes, it checks the heat of the iron and prevents its reaching the hand. In like manner the various iron implements that have to be thrust into the fire at one end and heated are furnished at the other end with a wooden handle by which they can be grasped without risk of burning the hand.

"Of all substances air is the poorest heat-conductor, as proved by the following curious experiment. A scientist named Rumford, to whom we are indebted for some noteworthy researches on the subject of heat, had some frozen cheese placed in the middle of a dish. Over this cheese was poured a light froth made of beaten eggs, and then the whole was put into a very hot oven to cook the eggs quickly. Thus was obtained a sort of *omelette soufflé*, piping hot, and in the middle of it was the frozen cheese, which had lost none of its coldness. How could the cheese remain frozen in the oven? The explanation is to be found in the poor conductivity of air. It was air, imprisoned in the froth of the eggs, that sheltered the cheese from the heat of the oven, obstructed the heat in its passage, and prevented its going farther. The heat not reaching it, the cheese in the center remained frozen."

"I should like," said Emile, "Ho have had a taste of that cheese, so cold under its hot crust of cooked eggs!"

"Now," resumed Uncle Paul, "I come to the practical application of the peculiarities I have just pointed out to you. A substance that conducts heat poorly may serve two purposes which at first seem contradictory and yet at bottom are alike in every respect. It can be used to protect both from the cold and from the heat, to keep an object warm as well as to keep it cool. To cool off is to lose heat; to get warm is to gain it. Accordingly, the point is to check, in the first case, the inner heat that might escape, and, in the second, the outer heat that might enter. Both requirements are met by the same means, the interposition of an obstacle impervious to heat in either direction; that is to say, a covering that is a very poor heat-conductor."

Then what keeps out the cold keeps out the heat too?" queried Marie.

"Precisely, singular though it may seem to you. The same covering that prevents loss of heat in any given body also prevents its receiving what might come from outside. A few examples will prove this to you. Let us first recall that chief among poor conductors are powdery and fibrous substances. This property they owe mainly to the air retained between one particle and another, one fiber and another, just as water is retained in the innumerable little cavities of a sponge. These substances are used as a protection from cold and heat alike. Ashes will furnish me my first example.

"If on going to bed at night we cover a bed of live coals with ashes, we shall find the coals still alive the next morning. The ashes, by keeping out the air, check combustion; but they do still more: at the same time that they check combustion they serve to retain in the coals nearly all their original heat, so that the next morning they are as glowing-hot as the night before. This result is due to the obstacle that ashes as a powdery substance oppose to the escape of heat. Under this powdery covering the embers remain alive because they cannot send out their heat, a poorly conducting body standing in the way.

"These same ashes that prevent the cooling of an object can also prevent its acquiring heat. I will tell you later how the little girls in my village used to go and borrow fire from a neighbor, returning with a live coal on a bed of ashes in the hollow of their hand. This shallow layer of ashes prevented their burning themselves, its poor conductivity arresting the heat of the glowing ember.

"The two examples are convincing: you see the same substance, ashes, acting as protector, first from the cold and then from the heat, keeping the live coals from cooling off and the hand from getting burnt."

Chapter Fifteen - Human Habitations

"I PASS to other examples. To keep the contents of a soup-tureen warm, what do we do? We enclose the tureen in a woolen coverlet; we wrap it in several thicknesses of a soft material which acts as a very poor heat-conductor because of its fibrous texture.

"But suppose, on the other hand, we wish to keep something at a low temperature; we again avail ourselves of this same quality possessed by fibrous substances. In summer, to exclude the heat from our ice-creams and sherbets we use a jar set into another of much greater size, filling the intervening space with wool, cotton, or rags. Thus the same wool, cotton, or other filling that maintains the warmth of the soup in the tureen will also serve to maintain the coldness of the frozen preparation in the jar.

"Ice, which is one of the most urgently needed supplies in warm countries, is sometimes transported long distances under a burning sun. The 'United States, for example, ships every year great quantities of ice to China and the Indies, and the vessels engaged in this traffic cross seas where the highest temperature prevails. Nevertheless the cargo reaches its destination almost intact, thanks to the non-conducting material protecting it from the heat, this material being sawdust, straw, dry leaves, or shavings, with which care is taken to cover completely the cakes of ice heaped up in the hold.

"An ice-house for keeping through even the hottest summer weather the ice gathered in winter consists of a deep excavation lined with brick in preference to stone, because the former is a much poorer heat conductor. A thick layer of straw is also placed next to the bricks. The ice is stored when the weather is very cold, the cakes being packed closely and then flushed with water, which freezes and renders the whole one compact mass. Over all is laid a bed of straw, and on top of this are put planks loaded with stones. Finally the ice-house is roofed with thatch. The very same procedure would be adopted if it were desired to keep from escaping the warmth present in the excavation."

"Then," said Marie, "this thatched roof and layer of straw would be just the thing for making a snug retreat that would keep out the cold."

"Yes, this snug retreat, made of straw and similar material to withstand the severities of winter, is exactly what we find in the extreme north of Europe, where the cold season is so rigorous. Houses of masonry like ours would there be inadequate, because brick and stone would offer but an imperfect obstacle to the dissipation of the interior warmth. For those arctic habitations some building material is required of less conducting power than brick and stone, some material that will retain the interior warmth just as ashes retain the heat of the live coals that they cover. To this end masonry gives place to walls of thick planks or even of logs laid lengthwise one upon another. So far so good, since wood conducts heat much less readily than stone; but it is not enough. A double enclosure or wall is built of these planks or logs, and the space between the two single walls is filled with moss, leaves, or straw. Thanks to this multiple enclosure made of materials that act as very poor heat-conductors, the warmth given out by a stove always alight is kept from escaping even when the severest cold prevails outside.

"The most curious application of powdery substances as a protection against cold is found in the use of snow for constructing winter dwellings."

"What," cried Claire, "do they build houses of snow?"

"Not exactly houses like ours, but huts in which the occupants are very well sheltered."

"And where is that?"

"In Greenland, to the northeast of America, in the frigid zone, as we have already seen. [1] There, where the rigors of the climate are well-nigh inconceivable to us, the Eskimos live in huts built of regular blocks of hardened snow laid in circular courses and rounding toward the top, with a sheet of ice capping the dome to admit the light. A bench of snow next to the wall encircles the interior and is provided with rude bedding of heather and reindeer skins for the repose of the inmates at night. No fireplace is ever found in these abodes, since there is no wood for fuel; and, moreover, a fire would melt the snow hut. A wick of moss fed with seal oil burns in a little stone pot and serves both to melt snow when water is needed for drinking and to maintain an endurable temperature in the hut, thanks to the very slight conductivity of the snow walls. Outside, meanwhile, the cold is of an intensity such that even our severest winters offer no comparison. If one leaves the hut, immediately face and hands turn blue and become quite numb; the skin cracks open under the action of the frost; the breath forms needles of hoar-frost around the nostrils, and the tears freeze on the edges of the eyelids."

"What a frightful country!" Claire exclaimed. "And do people really live there?"

"Yes, there are people who give the sweet name of home to that forbidding land. They live there the year round, in summer under skin tents, and when winter comes in snow huts."

"But why don't they build themselves houses of stone?" asked Jules.

"Because they would freeze in them," was the reply, "for lack of fuel to keep a hot fire going all the time. Snow is the only available material that can conserve the heat of the little lamp and maintain an endurable temperature in the hut; and this it does by reason of the poor conducting power of its powdery mass.

45

"The various powdery or fibrous substances, such as snow, ashes, sawdust, shavings, straw, moss, wool, cotton, feathers, all well adapted to the keeping out of either cold or heat, owe this peculiar property in great part to the air held prisoner in their interstices. The air imprisoned in the foam of beaten eggs prevents the heat of the oven from penetrating to the frozen cheese that I told you about in relating a curious experiment. The same foam, inflated with air, would protect from the cold any heated body that it might envelop. Air by itself can be used to prevent the dissipation of heat if it is so placed that it cannot be renewed and mix with the free air of the atmosphere. I will now call your attention to an instance in which this property of air is turned to account in our houses where the climate is severe.

"The warmth in a room escapes to the outside through the walls, floor, and ceiling, these never being perfect non-conductors. For this waste of heat there is hardly any remedy in our dwellings of brick and stone. But there is one avenue of escaping heat that can be easily closed. I refer to the windows. The panes of glass offer but a very imperfect obstacle to the outflow of heat. In order to supply a more efficacious barrier without lessening the transparency of the windows and their admission of light, it is customary to build, as it were, a wall of air behind the panes; that is to say, we fill the opening in the wall with two windows, one on the outside and the other on the inside, thus obtaining in the space between the two similarly glazed frames a layer of immobile air, a sort of transparent wall which the heat from within cannot pass through. That is what we call a double window."

[1] See "Our Humble Helpers."

Chapter Sixteen - Clothing

"WE say of one piece of goods that it is warm, of another that it is cool. What do we mean by that? Have fur and woolen cloth a heat of their own that they communicate to us? Do we demand of wool, down, or cotton some supplementary heat furnished by their own substance?

"Not at all; for none of these materials, be it the silkiest down or the softest fur, has any heat of its own, and they cannot supply it to us. Their office is limited to preventing the loss of such heat as is within us, the natural heat that our bodies generate simply by living. Garments and wraps are poor conductors interposed between the human body, warmed by the vital heat, and exterior objects which, colder than the body, would otherwise lower its temperature. They are to us what a shovelful of ashes is to the firebrands on the hearth. They do not warm us, but they preserve our natural heat; they do not add anything to us, but they prevent loss."

"Then a garment that we call warm, "said Marie, "is not really any warmer than another, but only keeps the heat of the body in better?"

"That is it, exactly. You see, apart from all idea of finery and pleasing adornment, the value of a garment from the sole point of view of real usefulness in retaining heat depends above all on its poor conducting power. The worse conductor it is, the better the garment will fill its office. In this respect wool heads the list of materials used in our textile goods; offering the most resistance to the radiation of the body's heat, it is the most serviceable in protecting us from

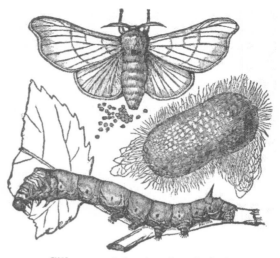

Silkworm (about natural size)

cold. Silk, cotton, hemp, and flax are far inferior to it. And that is as it should be. Wool is the sheep's clothing; therefore its attributes are assigned to it by Providence for protecting from cold the chilly animal that wears it. Silk, cotton, hemp, and flax have in the order of nature other uses: silk insures the safety of the caterpillar and its chrysalis in the cell of a firmly woven cocoon; the feathery down of the cotton-plant causes the seeds to float in the air and enables them to travel long distances and germinate in far-off places; the mission of hemp and flax is to strengthen long and fragile stems with their fibers. What was not clothing by nature became so by man's art, but it never acquires to the same degree the valuable qualities that belong to wool, the natural clothing of the animal it covers.

Cotton-Plant
Branch and (a) opened boll.

"Cotton in its turn conducts heat less readily than hemp or flax; hence cotton goods are preferable to linen for underwear that comes next to the skin and by its direct contact with us is so effective in maintaining our temperature. With cotton, sudden chills and arrested perspiration, with all their dangerous consequences, are less to be feared than with linen.

"But of all substances air is the poorest heat-conductor. Therefore we envelop ourselves, as it were, with air. Our textiles of wool, cotton, etc., are in a way only a sort of network for retaining the air in their innumerable meshes. This layer of air maintained all round our body protects us from cold the more effectively the thicker it is and the less liable to renewal. Hence it is not the heaviest and closest-woven fabric that is the warmest, but supple, soft stuff that receives an abundance of air throughout and holds it captive in its meshes, as do cotton-wool and down.

"Furthermore, between the body and its clothing, and retained by the latter, is found an envelop of air which we must take into consideration since it constitutes a natural lining that nothing could replace. To play its part well this lining of air must be of a certain thickness, which is assured by our wearing clothing that is loose enough but not too loose, because in the latter case the air, being renewed too easily, would be constantly replaced by cold air and, reversing its function, would cause a loss of heat instead of a gain."

"I must confess. Uncle Paul," Marie interrupted again, "I had no idea air played any such part in our clothing. Who would have imagined that we keep ourselves warm by holding a little stale air around us! Doubtless it is the same with our bed-coverings?"

"Precisely the same. Our bed-covers and hair mattresses are only barriers preventing the loss of natural heat. The light feathers, wool, cotton, hair composing them retain the air in abundance in their fluffy mass and thus form a non-conducting envelop which the heat of the body cannot pass through.

"Mattresses are even made of air alone without any fibrous material whatever. A substantial casing of canvas fastened in a frame is kept in shape by means of springs arranged inside. This pouch is filled with a layer of air, which is both soft and yielding, so as to insure comfort, and very effective in retaining heat. It is far preferable to heavy straw mattresses, being cleaner, less cumbersome, and easier to handle."

Chapter Seventeen – Ashes - Potash

"WASHING linen with ashes or lye is one of the most important operations of housekeeping. [1] In a large wooden tub the soiled linen is arranged with some care, on it is strewn a layer of wood-ashes, and over the whole is poured a quantity of hot water. This water, carrying with it the active constituents of the ashes, filters through the linen and removes the stains, running out continually in a small stream through an opening left in the bottom of the tub and collecting in a bucket from which it is drawn and replaced on the fire, to be poured over the ashes again when it is hot. The whole day is spent at this work. From morning to night always the same boiling water passes through the contents of the tub from top to bottom, cooling off on the way, running out, and returning to the fire to begin the same journey over again.

"Washing with the help of wood-ashes is much more effective than washing with hot water alone. The part played by the ashes is what we will now consider, and in this connection let us have recourse to a simple experiment.

"We drop a few handfuls of ashes into a pot of water and set it to boil. After boiling a little while we leave the contents to cool off. The ashes settle at the bottom and the water above becomes clear. Well, we shall detect in this liquid a peculiar odor in all respects like that which comes from the washtub; and we shall also find that it has an acrid, almost burning taste. This smell of lye and this acrid flavor were not in the water to begin with; they come from the ashes which have imparted a certain property not in the water originally.

"Ashes must therefore be composed of at least two elements differing from each other. The more abundant of these two cannot dissolve in water, but gathers in a mass at the bottom, forming an earthy layer; the other, on the contrary, constituting only a very small part of the whole, dissolves easily in water and gives to it its own peculiar properties, especially its odor and acridness. If, in order to make ourselves better acquainted with it, we wish to obtain by itself the part yielded by the ashes, nothing could be easier. All that is necessary is to put the clear liquid containing the solution in a vessel on the fire and heat it until all the water has evaporated. There will remain a very small amount of whitish matter looking a little like kitchen salt. But it is not kitchen salt, far from it; it is quickly recognized by its taste, which is unbearable. Put a little of this whitish powder on your tongue and you will instantly feel a prickly and painful sensation as if from a burn. The part touched would indeed be burned to the quick, just as if it had been seared with a red-hot iron, if the powder first underwent a certain preparation which I will not enter into now. The skin of the hand, although much less sensitive, is pained by prolonged contact with this harsh substance, which gnaws the skin and makes it crack and bleed. Wool, silk, feathers, hair, horn, leather, and almost everything of animal origin yield to its corrosive action and are at last reduced to a pulpy paste. Such is the active element in ashes, the element in fact that gives lye its drastic properties. It is called potash."

"Then washing with ashes produces its effect by means of the potash which the water dissolves and carries off through the layer of ashes?" This from Marie.

"That is it, exactly."

"Potash, which bites the tip of the tongue when one tastes it, and the skin of the fingers on handling it, also, you say, quickly destroys wool, silk, leather and many other things of animal origin. Then, woolen and silk goods should not be washed with ashes: the potash in the ashes would injure them."

"They would, in the end, go all to pieces in the wash."

"Hemp, cotton, and flax," continued Marie, "must be very tough not to be injured in a liquid capable of reducing wool to a pulp."

49

"I have already told you that textiles of these materials are endowed with an exceptional and admirable resistance which increases their value a hundredfold. Here you have a conclusive proof. A thick woolen stuff would come from the lye-wash all a sticky paste; a frail cotton fabric would emerge intact."

"Washerwomen's hands are all cracked," observed Claire. "I have seen some of these poor women with the skin worn off their fingers. It must be the potash from the lye that makes these wounds?"

"It is the potash. It eats into the hands as it would into a woolen stocking."

"Why, then," asked Emile, "do they put ashes in the wash-tub when this frightful drug, potash, destroys woolen things and tortures the washerwomen by eating into their hands? Why do they not simply use hot water?"

"That is exactly the point we are coming to. To get rid of an oil or grease spot what would you do, my dear child? Would you simply use water, hot or cold?"

"Certainly not. I know very well that water alone, even if boiling hot, would not take out the grease. I should use soap."

"Right. Well now, you must understand that if soap is good to take out grease spots it is just because it contains potash, as I will explain to you presently in detail. One more question: to wash dishes that are very dirty, very greasy, is hot water sufficient?"

"No; I think in that case ashes are boiled in the water, and with their help the grease comes off all right."

"Your answer is correct. Hot water alone cannot remove the grease, but hot water together with ashes does the work very well. In this case again the effect produced by the ashes is due to the potash they contain. This substance, in short, this potash that Emile calls a frightful drug, possesses a property very useful in housekeeping: it is the property of dissolving greasy substances of whatever sort they may be, whether oil, lard, suet, or tallow, and making it possible for them to be carried off by water. Try to take out with water alone the oil spot that soils a piece of linen: all your patience, all your efforts, will fail; the spot will remain afterward what it was before, and the water will have accomplished nothing. But if we first dissolve a pinch of potash in the water and then use it for washing, the spot will disappear without any trouble. To sum up, potash dissolves greasy substances and consequently gives water the power to take out spots produced by those substances.

"Now, more than half the stains on soiled linen belong to this class: they are grease spots. Prolonged contact with the human body leaves our garments full of impurities; little accidents at table soil the table-cloths and napkins with oil and grease; kitchen service accounts for all sorts of greasy deposits on the kitchen towels. To wash out these impurities on which water has no effect we are compelled to invoke the aid of potash, which is found in the ashes that lie ready to hand on the hearth. Ashes, then, play an indispensable part in the work of the laundry; with their energetic assistance hot wa-

50

ter is able to efface not only all grease spots, but also countless other stains that mere washing would not always remove.

"The ashes used are those that come either directly from wood or from wood that has been converted into charcoal. The best are those from bakers' ovens, on account of their larger proportion of potash. The dense wood of the trunk and larger branches of a tree contain less potash than the small branches and the leaves. Hence the fuel used for heating ovens, namely bundles of fagots, yields better ashes than we commonly find on our hearths. Finally let me add that coal-ashes are absolutely valueless and would even be injurious."

[1] This method of washing, still retained in certain rural districts of Europe and not unknown in America, has now largely given place to quicker and easier processes. — *Translator.*

Chapter Eighteen - Soap

"The ashes of plants that grow on the seashore and in the waters of the sea itself contain, instead of potash, another substance possessing almost the same properties and called soda."

"There are plants, then, in the sea?" asked Jules.

"Certainly, my boy; and most curious ones. Our fields are not more covered with plants than is the bottom of the sea; but marine plants differ much from terrestrial ones. Never do they have flowers, never anything to compare with leaves, never any roots; they fasten themselves to the rock by gluing the base of the stem to it, d but derive from it no sustenance whatever. It is water and not the soil that nourishes them. There are some that resemble viscous things, plaited ribbons, or long manes; others that take the form of little tufted bushes, of downy topknots, or of waving plumes; there are some of pinked strips, others coiled in a spiral, and still others fashioned like coarse, slimy strings. Some are olive-green, some pale rose; others are

Seaweeds (*Algæ*)

1, *Dictyota dichotoma*: *a*, spore; *b*, vertical view of a cystocarp; *c*, vertical section of same. 2, *Plocamium coccineum*: *f*, tetraspore; *g*, stichidium; *h*, branchlet with a cystocarp.

yellow like honey, and others, again, of a bright red. These strange plants are called algae, or seaweeds.

"For a long time the only way to obtain soda was to gather such marine plants as were cast upon the shore by the waves and to cut down the various

51

kinds growing at the water's edge. As soon as this material was quite dry it was burned out of doors in a ditch, so as to get the ashes. But as this method was slow and did not furnish enough soda to supply what is needed in the arts, which use an enormous quantity, men of science set their wits to work to devise more fruitful and more expeditious methods. To-day soda is prepared from common salt, the sea furnishing inexhaustible quantities. Immense factories, each with a large force of workmen, are occupied exclusively with this work."

"Soda, then, is something very important!" asked Claire.

"Yes, one of the most important articles used in manufacture. Many things of the greatest utility and in everyday demand need for their fabrication the cooperation of soda. The fine, white paper on which you write, the magnificent colored designs of our calicoes and other prints, the glass in our bottles and window-panes, our soap, that invaluable aid to cleanliness, all these things and countless others require the use of soda, or of potash, which I will call its sister, so much do they resemble each other."

"Then potash, too, must be manufactured on a large scale?" was Marie's query.

"The manufacture of potash stands on a par, in importance, with that of soda; but potash is always dearer because less abundant. The greater part of it is obtained from the ashes of terrestrial plants. For this purpose in well-wooded countries such as certain regions of Russia and North America, whole forests are cut down and the wood is burned on the spot, out in the open air, just for the sake of the ashes."

"Do you mean to say," said Marie, incredulously, "that those enormous fires that consume whole forests are lighted for nothing but to get the ashes?"

"For nothing but to get ashes from which to extract the potash. I hasten to add that this prodigality is possible only in countries where forests are abundant and the population very sparse. In such regions wood is of little value, as there are no people to use it for heating. But in our part of the world, where the forests are far from sufficient for the needs of heating, great care is taken not to waste wood thus. In the forest-covered mountains of the Vosges, for example, it is the practice, in order to obtain ashes, to burn only the very small branches, which are of little value, and the dead leaves.

"The manufacture of potash is conducted in the manner I showed you a little while ago in our elementary experiment. First the ashes are boiled in water, and then the clear liquid is drawn off and evaporated to the last drop over a fire. The incrustation at the bottom of the kettle is the potash, which is further purified by methods that do not concern us here.

"The properties of these two valuable substances, soda and potash, are nearly identical. Both have a beautiful white color when very pure; at first sight you would take them, in their unpowdered form, for white marble. Both dissolve very easily in water, to which they give the odor and taste of lye. Both have a very disagreeable taste; a tiny piece, less than a pin's head, placed on the tongue, as I have already told you, would burn like a red-hot

52

iron and take off a piece of the skin. Both eat into leather, wool, and silk; and both dissolve greasy substances. You have seen this last attribute demonstrated in the use of ashes in washing; and the same characteristic is found again in soap, as I will now explain to you.

"Of all the stains that are left upon linen by daily use, the most frequent, as you know, are grease spots, which water alone cannot remove. To take out these spots they must first be rendered soluble in water by adding a substance which will give them solubility. Potash and soda fulfil this condition admirably. But direct use of these harsh substances is impracticable. What would become of the washerwoman's hands, rubbing the clothes with drugs that burn the skin like fire I In a few moments they would be nothing but one horrible wound. And that is not all: the linen itself, however strong it might be, would finally be destroyed by such prolonged contact with these excessively drastic substances. Potash and soda, therefore, can in no wise be employed directly in washing. What is to be done then? The difficulty is solved by adding another substance which takes away their formidable strength without too greatly weakening their solvent effect. To temper the excessive energy of potash and soda, to soften in some measure the two terrible drugs and make them easy to handle, they are mixed with a greasy ingredient, sometimes oil, sometimes tallow; and from this mixtures comes soap."

"There is oil or tallow in soap?" asked Emile.

"Yes, my dear boy, and plenty of it. The rest consists of a little potash or soda. These latter give soap the power to clean; the oil and tallow preserve the hands and the linen from a contact which without any intermediary substance would be very dangerous."

"And yet, apart from the greasy feeling, there is nothing in soap to show that it has any tallow, still less oil. Oil runs, and soap does not."

"Oil runs only when alone. Once united with potash or soda, it ceases to be liquid and becomes a block of the consistency of cheese. Singular though it may seem to you, it is nevertheless true that soap is composed of either soda or potash and a greasy substance, oil or tallow.

"Soap for common use is made from soda and oil of inferior quality, or beef suet or mutton tallow. Let me tell you how soap in large quantities is made. Into great vats of boiling water the desired quantity of soda is dropped, then the due proportion of oil or grease, after which the whole is stirred constantly to mix it well together. Little by little the soda becomes incorporated with the grease, and soap forms and floats on the surface in a compact layer, which is then taken off and poured into molds where it congeals in thick square slabs. These slabs are afterward divided into cakes of a convenient size.

"There are two kinds of ordinary soap, white soap and marbled soap. The first is white throughout, the second veined with bluish lines. For common use marbled soap is preferable to the other. Besides soda and grease or oil, soap of any kind always contains a greater or less proportion of moisture coming from the water in which it was boiled. Now, white soap contains nearly half of its weight in moisture, while marbled soap contains not quite a third. Being richer in the ingredients that really count, marbled soap is for that reason more economical.

"Resinous soap is a kind of soap that contains resin instead of tallow or oil. This soap is of a yellow wax color, and its cakes are transparent on the edges. It makes a great deal of lather on being dissolved in water, is very strong, and is good for washing clothes.

"Toilet soap is prepared from choice materials and is perfumed with various aromatics embodied in its substance."

Chapter Nineteen - Fire

"WE do not know how man first procured fire. Did he take advantage of some blaze started by a thunderbolt, or did he kindle his first firebrand in the crater of a volcano? No one can tell. Whatever may have been its source, man has enjoyed the use of fire from the earliest times; but as the means of relighting it if it went out were very imperfect or even lacking altogether, the utmost care was taken to maintain it, and a few live coals were always kept over from one day to the next.

"So calamitous would have been the simultaneous extinction of the fires in all the dwellings that, in order to guard against such a disaster, the priesthood took fire under its special protection. In ancient Rome, many centuries ago, an order of priestesses called Vestals was charged with the guarding of the sacred fire night and day. The unfortunate one who let it go out was punished with horrible torture: she was buried alive!"

"Did they really bury her alive for letting the fire go out?" asked Jules.

"Yes, my boy. This terrible punishment inflicted on the keepers of the fire shows you the importance they attached to keeping at least one hearth alight so that others could be kindled from it."

"One of our matches that we buy at a cent a hundred," said Claire, "would have saved the life of the careless Vestal."

"Yes, to abolish those barbarous severities it needed only a match, a thing which unfortunately was at that time unknown.

"Many centuries passed before it was discovered how to procure fire easily. In my young days, when I was about your age, keeping coals alive to be used for relighting the fire next day was still the rule in the country. In the evening before the family went to bed, the embers were carefully covered with hot ashes to prevent their burning out and to keep them alive. If, despite this precaution, the hearth was cold next morning, some one had to hasten to the nearest neighbor's to borrow some fire, that is to say a few live coals, which were carried home in an old wooden shoe to keep the wind from blowing them away."

54

"But I should think the old wooden shoe would have caught fire," said Emile.

"No, for care was taken to put a layer of ashes in first. I have told you how some children would put a few ashes in the hollow of their hand, and on the ashes lay live coals. They carried fire thus just as you would carry a handful of sugar-plums.

"The layer of ashes arrested the heat of the embers and prevented its reaching the hand. Remember what I have already told you about the poor conducting power of ashes, their refusal to transmit heat, a characteristic they have in common with all powdery substances. The little fire-borrowers knew that well enough,"

"But who taught them to do it that way! "asked Emile.

"The great teacher of all things, necessity. Caught without shovel or wooden shoe, some one of them, knowing this peculiarity of ashes in arresting heat, made use of the ingenious device I have described, and his example was sooner or later followed by others.

"Fire-producing devices are, as a rule, based on the principle that heat is generated by friction. We all know that we can warm our hands by rubbing them against each other."

"That's what I always do in winter when my hands are frozen from making snowballs," said Jules.

"That is one of the oldest illustrations of the effect of friction, and I will add another. Hold this round-headed metal button by the shank and rub it briskly on the wood of the table; it will become warm enough to produce a decided feeling on the skin."

Claire took the button, rubbed it on the wood of the table, and then applied it quickly to her hand, uttering a little cry of surprise and even of pain as she did so.

"Oh, how hot the button is. Uncle! "she exclaimed. "If I had rubbed any longer I should have scorched my hand."

"It is by similar means that certain savage tribes procured and still procure fire. They twirl very rapidly between their hands a slender stick of hard wood with its pointed end inserted in a cavity hollowed in soft and very inflammable wood. If the friction is brisk enough and the operation properly carried out, the soft wood catches fire. This process, I admit, would fail of success in our hands for lack of skill."

"For my part," said Marie, "if I had nothing but a pointed stick and a piece of wood with a hole in it for lighting a fire, I should despair of ever managing it."

"I should not even try it," Claire confessed, "it seems so difficult, although the button that I rubbed came near burning me."

"What would be impossible for us is mere play for the natives of Australia. The operator sits on the ground, holding between his feet the piece of wood with the little hole, and twirling the pointed stick rapidly between his hands he soon obtains a spark with which he kindles a few dry leaves.

55

"Even in our own country you may see, in any wood-turner's shop, this friction process employed successfully. To obtain the brown ornamental lines on certain objects turned in a lathe, the operator presses with some force the point of a bit of wood on the piece in rapid rotation. The line thus impressed by friction begins to smoke in a few moments, and soon becomes carbonized.

"I pass on to other methods of producing fire. Iron and steel, especially the latter, if rubbed against a very hard stone give out sparks made by tiny scales of metal that become detached and are sufficiently heated to turn red and burn in the air. Thus the scissors-grinder's revolving stone, although constantly moistened with water, throws out a shower of sparks under the steel knife or other tool that is being sharpened. In like manner the cobblestone struck by the horse's iron shoe emits sudden and brilliant flashes.

"The common flint-and-steel apparatus acts in the same way. It consists of a piece of steel that is struck against the edge of a very hard stone called silex or flint. Particles of steel are detached from the metal and, made red-hot by the friction, set fire to the tinder. This latter is a very combustible substance obtained by cutting a large mushroom into thin slices and drying them, the mushroom being of the kind known as touchwood, which grows on tree trunks."

Chapter Twenty - Matches

"As tinder burns without flame, the glowing spark obtained with flint and steel before the invention of our matches did not suffice for obtaining fire; we had to have recourse to sulphur, which possesses the invaluable property of bursting into flame at the mere touch of a red-hot substance.

"Sulphur is so well known to you as to render any description of it here unnecessary. It is found especially in the neighborhood of volcanoes, where it occurs in the soil, sometimes in masses free from all admixture, sometimes in mingled masses of sulphur and earth or stone. Man's work consists merely in purifying the sulphur by melting it as it comes from the mine.

"In the olden time matches were made of pieces of hemp dipped at one end into sulphur. They were lighted by having the sulphur-tipped end touched either to a live coal that was kept glowing under the ashes, or to a bit of tinder previously kindled by flint and steel. Thus you see the mere lighting of a lamp was a process not devoid of complications. First the flint and steel must be struck together, at the risk of bruising one's fingers by an awkward movement in the dark; then, when the tinder had taken fire after many attempts which often exhausted the patience, it was necessary to apply the sulphur match in order to obtain a flame."

"Our matches of to-day are much to be preferred," remarked Claire. "All you have to do is to strike them against the box cover or against the wall or a piece of wood, no matter where, and the thing is done: the fire burns."

"This inestimable benefit of being able to obtain fire without difficulty and on the instant we owe to phosphorus, a substance discovered, as I have already told you, [1] by a learned investigator named Brandt, who lived in Hamburg two hundred years ago. In attempting the impossible transformation of baser metals into gold, he hit upon an elementary substance until then unknown, and thus gave us the self-igniting sulphur match with its tip of phosphorus.

"If you examine one of our common matches you will see that the inflammable end is coated, first with sulphur, and then, over this, with phosphorus, the latter being colored with a red, blue, or brown powder, according to the maker's fancy. Phosphorus by itself is yellowish and translucent like wax. Its name means light-bearer. When it is rubbed gently between the fingers in a dark place, the fingers are seen to be covered with a pale light. At the same time a smell like that of garlic is detected; it is the odor of phosphorus. So inflammable is this substance that it takes fire when heated only a very little or when rubbed against any hard surface. Hence its use in the manufacture of friction-matches.

"These are little sticks of wood — willow, poplar, or spruce — wrought with the help of steel plates pierced with holes having sharp cutting edges through which the wood is forced by powerful pressure. Then these little sticks, held in position by frames made for the purpose, are first dipped at one end into melted sulphur. Over this first coating, which is designed to feed the flame and give it sufficient intensity to ignite the wood, must be laid a second that will take fire by friction; and this latter coating is composed chiefly of phosphorus. On a marble table is spread a semi-fluid paste made of phosphorus, glue, very fine sand, and some coloring matter. The matches, still held in position by the frames just referred to, have their sulphur tips touched for an instant to this inflammable paste, and are then placed in an oven where the paste is allowed to dry. Friction, aided by the fine sand incorporated in the paste, develops enough heat to ignite the phosphorus; this in turn sets the sulphur on fire, and from the sulphur the flame spreads to the wood.

"Phosphorus is a deadly poison, and therefore matches must be handled with care, this precaution extending even to the empty boxes that have held them. Contact with our food might entail serious consequences. Nevertheless this fearful substance is found in all animal bodies. It is present in the urine, whence Brandt was the first to extract it; it occurs in meat, in milk, and above all in bones. Plants also, especially cereals, contain it, and hence it enters into the composition of flour and of bread."

"What!" exclaimed Claire, "a substance so frightful, a poison so violent, is found in milk, meat, and bread?"

"Don't be alarmed," her uncle reassured her. "We run no risk whatever of being poisoned by drinking a glass of milk or eating meat and bread. The phosphorus there present does not occur by itself, but combined with other substances which deprive it entirely of all poisonous attributes and render it useful, in fact necessary, to the strength of the body. It is to be feared as a poison only in the condition in which it is found in matches. I should add in conclusion that the method adopted by Brandt for obtaining phosphorus — namely, from urine — has long since been abandoned. At the present time it is extracted from the bones of animals."

"Then it is bones that furnish us with phosphorus?" said Jules.

"Yes, bones are by an ingenious device made to yield us phosphorus and, consequently, light and heat."

[1] See "Field, Forest, and Farm."

Chapter Twenty-One - Wood and Charcoal

"OF wood used for fuel it is customary to distinguish two kinds, hard and soft. At the head of the former class stand the different species of oak, notably the common oak scattered all over France, and the evergreen or holm oak peculiar to the south. This last is called evergreen because it does not shed its foliage in winter, but continues green the year through. Among the soft woods are the poplar, willow, plane-tree, and pine.

"The way in which these two kinds of wood burn is quite different in the two instances. Soft wood, suitably dried, takes fire readily and gives much flame with a heat that is quick but of short duration. It is convenient for use in the kitchen, where it is often necessary to obtain a prompt and intense heat, for example in roasting fowl on the spit and in cooking with the frying-pan. Furthermore, soft wood is very useful for kindling other and less combustible fuel, such as hard wood, coal, and charcoal. But for use in open fire-places, when it is desired to keep a fire going all day long, soft wood is by no means economical, because it bums up so quickly. The best fuel in this case is oak, which bums slowly and yields a large, compact mass of coals that retain their heat for hours at a time, especially if care is taken to cover them partly with ashes."

"I have noticed the difference you speak of," said Marie. "Willow and poplar burn to ashes in a few minutes, leaving hardly any coals; but oak gives a fire that lasts and at the same time leaves a mass of glowing coals."

"If we had to keep up a fire with dry twigs and willow splinters," remarked Claire, "we should be kept busy all day throwing on wood, whereas three or four sticks of oak last a long time."

Emile here interposed with a question. "How is charcoal made," he asked, "such as is used for cooking."

"Charcoal is made from wood," was the reply. "Its superiority to wood is that it bums almost without smoke and without flame, a very valuable quality where cleanliness is desired, as it is in our kitchens. More than that, charcoal yields a heat that is equable and lasting, thus dispensing with the need of careful watching. The best charcoal is obtained from the best wood, that is to say from oak, especially evergreen oak. I will now describe its mode of manufacture as followed by charcoal-burners in the heart of a forest.

"Upon a plot of ground beaten hard and level there is first built a sort of chimney made of logs planted vertically in the soil, and around this chimney the wood destined to be converted into charcoal is piled in tiers, one on top of another, but with openings left at the base for admitting air. The whole is covered with a layer of earth and sod, leaving exposed only the central chimney and the air-holes at the base. Finally, the mass is set on fire by means of dry brushwood."

"I should think," said Jules, "that the whole pile would burn up and leave nothing but ashes, a dead loss to all concerned."

"By no means," replied Uncle Paul. "Since air can reach the burning mass only with difficulty, combustion is slow and the wood is but half consumed. Besides, if the fire should burn too briskly the attendants would make haste to stop up with sod some or, if necessary, all of the vents at the base of the pile. As soon as the pile is thought' to be one mass of glowing coals, the fire is smothered with earth and the structure is left to cool off. When this process is complete the work of demolition is undertaken, and in place of wood in its primitive state there is found nothing but charcoal. Some few fragments there may be that are not thoroughly charred, and they are recognized by their reddish color. They are the half-burnt pieces we find now and then in our charcoal."

"I know them," said Claire, "They make such a disagreeable smoke; and I always take them out of the stove as quickly as I can and throw them into the fireplace."

"That is right," returned Uncle Paul. "And now let us direct our attention to another kind of coal, the coal that is dug out of the earth and that comes to us from the coal-mines. Coal of this sort is of vegetable origin, no less than the charcoal whose mode of manufacture in the heart of the forest you now understand. But charcoal-burners do not make this other coal; it is found all made in the bowels of the earth, at great depths below the surface."

"But how," asked Claire, "can coal from deep down in the earth come from vegetable matter growing only on the surface."

"To explain that to you in full is out of the question, because your knowledge is still too elementary; but I can at least give you some idea of the natural processes involved.

"Let us suppose the existence of great forests of luxuriant growth, forests that man can never penetrate with his instruments of destruction. The trees fall under the weight of years and go to decay at the foot of other and younger trees, forming at last a layer of matter half carbonized by the action of the elements. One generation of trees succeeds another, and the layer grows thicker, so that after centuries it attains the thickness of a meter or more. Imagine now a succession of violent earthquakes which break up the surface of the earth, pushing up mountains where before there have been plains, and making plains where there have been mountains. Imagine, further, that as a result of these changes of level the sea is displaced and forced to forsake, wholly or partly, its old bed for a new one; conceive of this new sea as covering the shattered remains of old forests with mud and sand that eventually become hardened and converted into thick beds of rock; and, finally, picture to yourselves the sea as at last, in consequence of still further upheavals of the earth's surface, forsaking the new bed it had found for itself and seeking still another, leaving behind it a continent of dry land. Thus you will have all the essential facts that you require in order to understand the presence of coal in the interior of the earth."

"But those frightful upheavals that you speak of," put in Jules, "those great changes that make continents of seas and seas of continents — have they ever really taken place?"

"A science called geology," replied Uncle Paul, "teaches us that these events actually occurred as I have described them, but so long ago that man was not yet in existence.

"There was a time, for example, when this comer of the globe that bears to-day the fair name of France was composed merely of a few small islands lost in a vast ocean. On these islets, which were covered with lakes and volcanoes, there flourished a luxuriant vegetation, the like of which is no longer anywhere to be found except perhaps in the depths of some tropical jungle. The very regions where now grow beech trees and oaks used to bear immense ferns, each balancing at the top of its tall stem a graceful cluster of enormous leaves. These fern trees constituted the greater part of somber, damp forests which were never enlivened by the song of birds and never heard the step of quadruped; for as yet the dry land was without inhabitants. The sea alone maintained beneath its billows a population of monsters, half fish, half reptile in form, their flanks clothed in an armor of glittering scales. The remains of that ancient vegetation, buried in the depths of the earth by some stupendous cataclysm, have become the coal-beds of to-day in which are still discerned the admirably preserved outlines of leaf and stalk."

Chapter Twenty-Two - Coal and Coal-Gas

"IF I should take it into my head," Uncle Paul resumed, "to begin my talk to-day in some such fashion as this, 'Once upon a time there lived a king and a queen,' and if I introduced into my fanciful tale a wicked ogre greedy for human flesh and a good fairy borne over the surface of the sea in a mother-of-pearl shell drawn by red fishes, I am firmly convinced that your attention would at once be riveted to my account of the deeds done by king and ogre and fairy, so that you would give me no peace until the story was finished.

"Now, real things can offer to your curiosity food that is just as wonderful and that, moreover, feeds the mind with useful knowledge. Coal, the history of which I sketched in outline at our last talk, is one example among a thousand."

"It is very true. Uncle," said Marie, "that Cinderella or Bluebeard never interested me as does the story you began to tell us about coal. Those little islands that afterward became France by changes in the ocean-bed, those ancient forests which man has never seen, but which to-day he finds transformed into coal deep down in the earth, those upheavals in the surface of the earth, turning everything topsy-turvy, all excite in me the liveliest interest, and I should very much like to hear more about them."

"Let us talk a little more, then, about coal. But what could I select humbler in appearance and less worthy of your attention than that black stone! Nevertheless there are some most astonishing things to be said about it. First of all, coal — black, dirty, with no claim to distinction in its aspect — is own brother to the diamond, the sumptuous gem that for brilliance has no equal in all the world. The diamond is made of carbon, and so is coal — the very same substance in both."

"The diamond made of carbon? "cried Claire, incredulously.

"Yes, my child, of carbon and nothing else."

"But-"

"There are no buts. Again I say the diamond is made of carbon, but in a perfectly pure and crystalline form, which explains its transparency and brilliance. I told you that coal had things to tell us of a most marvelous nature."

"I see now that it has," assented Claire.

"Furthermore, this piece of black stone traces its origin to remote ages when it made part of some elegant tree such as you will nowhere find at present unless perhaps in some tropical region enjoying an abundance of heat and moisture."

"Then it must have come from one of those ancient forests you told us about yesterday," said Emile.

"Yes, and I can prove it to you. As a rule coal occurs in shapeless masses which furnish no indication of their origin; but not infrequently there are found, in the cleavage between two layers, distinct traces of carbonized vegetable matter having perfectly recognizable outlines. Certain coal-beds are formed entirely of leaves heaped up and pressed together into a solid block, and still preserving, despite their conversion into coal, all the details of their delicate structure. These relics of a plant life as old as the world, wonderful archives which tell us the earth's history, are so well preserved that they enable us to recognize the carbonized plants with the same certainty we feel in recognizing living plants. Yesterday, while our winter supply of coal was being put in, I chanced to notice in the bin some of those venerable relics I have just been speaking about, and I put them aside to show to you. Here they are."

"Oh, what beautiful leaves!" exclaimed Marie. "How nicely they are attached to the black surface underneath! One would say they had been cut out of very thin sheets of coal." She stood lost in thought before these remains of forests so extremely old as to antedate all animal life.

"When the plant life was flourishing which you see represented here," went on Uncle Paul, "The earth was covered with a vigorous growth of vegetation unexampled in our own time. This vegetation, buried far underground by changes in the earth's surface, and carbonized in the course of a long series of, centuries, has become transformed into enormous masses of coal which constitute the soul, so to speak, of modern industry. For it is coal that moves the railway locomotive, with its line of heavy cars; it is coal that feeds the furnaces of factories; it is coal that enables the steamboat to brave wind and storm; and it is coal that makes it possible for us to work the various metals and manufacture our tools and instruments, our cloth and pottery, our glassware and all the infinite variety of objects necessary to our welfare. Are you not filled with wonder, my children, as I am, that long before man's appearance upon earth everything was prepared for his reception and for providing him with the things essential to his future industry, his activity, his intelligence! Are you not impressed by this vegetation of prehistoric times which stored up in the bowels of the earth those precious deposits of coal that to-day are brought to light and made to move our machines and become one of the most active agents of civilization?"

"From now on," replied Marie, "whenever I put a shovelful of coal on to the fire, I shall think of that ancient plant life which gave us this fuel."

"Nor is the whole story of coal told yet," Uncle Paul went on. "Besides heat, coal gives us light. Cities are illuminated by street lamps which bum no oil and have no wick, but emit a simple jet of gas which, on being ignited, produces a magnificent white flame. [1] This gas is obtained by heating coal red-hot in great air-tight ovens. Pipes laid under ground conduct the gas from the gas-works to all parts of the city and distribute it to the street lamps. At nightfall the burner is opened and the gas flows, taking fire at a little hand-lantern with exposed wick, whereupon the flame bursts forth.

"What remains in the ovens after the manufacture of illuminating gas is a modification of coal known as coke, an iron-gray substance with a dull metallic luster. Coke develops much more heat than the best wood-charcoal, but is difficult of combustion and in order to burn well must be heaped up in considerable quantity and have a good draft. For domestic heating it is used in stoves and, still oftener, in grates. It is superior to coal in giving forth no smoke, thus being cleaner.

"Together with gas there is produced in the ovens in which coal is heated a black and sticky substance called coal-tar. From this horrible pitch, which one cannot touch without soiling one's hands, modern invention knows how to extract something in the highest degree fresh and beautiful and fair to look upon. As we have seen, the splendid colors of our silks and cottons, the rich and varied tints of our ribbons — all these we owe to dyestuffs obtained from coal-tar. Common coal, therefore, far from impressive though it is in appearance, is linked with the most dazzling splendors this world can produce: on the one hand with the diamond, with which it is one in essence, and on the other with the flowers of the field, whose delicate coloring it imitates and even surpasses."

[1] Only older readers will recall this method of street-lighting, which has long since been superseded by electric lamps of various kinds, — *Translator.*

Chapter Twenty-Three - Combustion

"LET us light a shovelful of charcoal in the kitchen stove. The charcoal catches fire, turns red, and is consumed, while at the same time producing heat. Before many minutes there is nothing left but a handful of ashes weighing but a trifle compared with the quantity of charcoal burned. What, then, has become of the charcoal?"

"It has been consumed," answered Jules; "it is burnt up."

"Agreed. But being consumed — does that mean being reduced to nothing? Does charcoal, when once it has been burned, become nothing at all, absolutely nothing?"

"It has turned to ashes," Jules replied.

"You haven't hit it yet, for the ashes left after combustion amount to very little compared with the quantity of charcoal consumed."

"Your question, dear Uncle," Marie here interposed, "puzzles not only Jules, but me too, very much indeed. If there isn't any more than a handful of ashes left after the charcoal is burnt up, I should say the rest has been destroyed."

"If that is your opinion I would have you know that in this world nothing is ever entirely destroyed, not a particle of matter ever becomes nothing after having been something. Try to annihilate a grain of sand. You can crush it, convert it into impalpable powder; but reduce it to nothing — never. Nor could the most skilful of men, men equipped with more varied and more powerful appliances than ours, succeed any better. In defiance of every exertion of ingenuity or violence the grain of sand will still continue to exist in some form or other. Annihilation and accident, two big words that we use at every turn, really do not mean anything. Everything obeys laws: everything persists, is indestructible. The shape, aspect, appearance, changes; the underlying substance remains the same.

"So the charcoal that is burned is not annihilated. True, it is no longer in the stove; but it is in the air, dissipated and invisible. When you put a lump of sugar into water the sugar melts, is disseminated throughout the liquid, and from that time ceases to be visible even to the keenest scrutiny. But that sugar has not ceased to exist. The proof is that it communicates to the water a new property, a sweet taste. Furthermore, nothing stands in the way of its ultimate reappearance in its original form. All we have to do is to expose the sweetened water to the sun in a saucer; the water will disappear in vapor and the sugar remain.

63

"Charcoal behaves in like manner. In burning it is dissipated in the air and becomes invisible. This dissipation is called combustion. What do we do when we wish to make the fire burn faster? With the bellows we blow air on the fuel. With each puff the fire revives and burns brighter. The live coals, at first dull red, become bright red and then white-hot. Air breathes new life into the bosom of the burning mass. If we wish, on the other hand, to prevent a too rapid consumption of fuel, what do we do! We cover the fire with ashes, thus keeping out the air. Under this layer of ashes the live coals retain their heat and remain red for a long time without being consumed. Thus it is that a fire in a grate is maintained only by the constant admission of air, which makes the coal burn; and the faster it burns, the greater the amount of heat given off."

"Then that must be why the stove gets so hot when it roars," remarked Claire. "Air is let in between the bars of the grate and then goes roaring through the red-hot coals. But if the air is prevented from circulating by closing the door of the ash-pit, the heat subsides at once."

"When the air is impregnated with carbon it acquires new properties, just as water does on becoming charged with salt or sugar. This new element is an injurious substance, a harmful gas, all the more to be feared because it does not reveal its presence, having neither smell nor color. We do not take note of it any more than we do of ordinary air.

"But let any one breathe this formidable gas, and immediately the brain becomes clouded, torpor supervenes, strength fails, and unless help arrives death soon follows. You have all heard of unfortunate persons who inadvertently — sometimes, alas! designedly — have met death in a closed room by lighting a brazier. The fact that the air becomes impregnated with the dissipated carbon from the burning charcoal explains these lamentable accidents. Inhaled even in a small quantity, this deadly gas induces first a violent headache and a general sense of discomfort, then loss of feeling, vertigo, nausea, and extreme weakness. If this state continues even a very little while, life itself is endangered.

"You see to what a risk charcoal exposes us when the products of its combustion do not escape outside through a chimney, but spread freely indoors, especially if the room is small and tightly closed. Under these conditions you cannot be too distrustful of a brazier. Whether burning brightly or half extinguished, whether covered with ashes or not, these embers exhale a deadly gas which does not announce its presence by any sign that we can recognize, but, like a traitor, always takes us by surprise. Death may occur even before danger is suspected.

"Again, it is very imprudent to close the damper of a bedroom stove for the sake of maintaining a moderate heat during the night. The stovepipe being closed by the damper, there is no longer any outlet for the products of combustion, which are sent out into the room and asphyxiate the sleepers, so that they pass from life to death without even waking up. [1]

"If the apartment is small and without openings for changing the air, a simple foot-warmer is enough to cause a headache and even lead to more serious results."

"Now I understand," said Marie, "The headaches I sometimes have in winter when I am sewing in my little room, all shut up with a foot-warmer under my feet. It was the burning charcoal that gave me those headaches. That is a good thing to know, and I will be careful in future."

"Be just as careful with charcoal when you are ironing. Keep the heater for the irons under the chimney or in a well-ventilated place, so that the dangerous exhalations from the embers may be carried out into the open air. Those who do ironing often complain of an uncomfortable feeling which they attribute to the smell of the iron, whereas it is due to the deleterious gas given off by the burning charcoal. It can be avoided by keeping the heaters under a chimney or in a current of air that drives away the injurious gas."

[1] With proper ventilation, as the author might have added, this danger is greatly lessened. — *Translator.*

Chapter Twenty-Four - Heating

"I HAVE told you how man has possessed fire ever since the earliest times. The first fireplaces for preparing food and furnishing protection from cold consisted of armfuls of fagots burning between two stones, either in the open air or in the middle of the hut. This rude method of domestic heating still prevails among many savage tribes. On flagstones in the middle of the dwelling smolder a few firebrands, the smoke escaping as best it can through some chance cracks and crannies in the roof. Indeed, if you wish to view this primitive method of making a fire, you need not go to distant countries beyond the reach of the benefits of civilization. In certain mountain cantons of France the fireplace is still to be seen in the form of a large flat stone in the middle of the room, the walls and rustic furnishings of which have become coated by the action of smoke with a brilliant Varnish as black as shoe-polish. The cracks in an imperfectly fitted roof furnish the only outlet for the products of combustion."

"The family must be nearly smothered to death," said Claire, "when the fire smokes in the middle of one of those chimneyless houses. Why don't they make a fire like ours?"

"Chimneys are a rather recent invention, and in those remote mountain villages habit preserves old traditions indefinitely. Antiquity in its period of highest refinement knew absolutely nothing of our common chimney. A striking proof of this is given us by a celebrated city, Pompeii, which was buried in the year seventy-nine of our era under a bed of volcanic ashes thrown up by Mount Vesuvius. Its houses, after having been buried for eighteen centuries, are exhumed to-day by the miner's pick and come to light again as they were when overwhelmed by the volcano. Not one of them has a chimney. In the old Roman town people warmed themselves with burning charcoal placed in a large metal vessel on a bed of ashes. This portable fireplace was put in the middle of the room to be warmed, without the slightest reference to air currents or the escape of harmful gases engendered by the burning charcoal. And even in our own time, in Italy and Spain, similar open braziers are used."

"I should think," said Marie, "that there would be danger of accidents, or at least that the people would have headaches due to the burning charcoal."

"The mildness of the climate, which does not call for air-tight houses, permits this vicious mode of heating in Spain and Italy; but braziers would be very dangerous in our homes, where windows and doors must be carefully closed during the winter. Unchanged air impregnated with the deleterious gases from combustion would soon lead to discomfort and even asthma.

"The first chimney-places for domestic use mentioned in history date from the fourteenth century. Disproportionately large, very costly, burning whole trunks of trees for fuel, these chimney-places were at first constructed without any knowledge of how to economize heat. Immense fires were built, but with no resulting warmth proportioned to the fuel used. The subject of draft was not in the least understood, and it was not until the end of the last century that there was any clear perception of the truth that draft in a fireplace is caused by the difference in temperature between the air of the chimney-place and that outside.

"I will now call your attention to something you have witnessed a thousand times in winter when you sit around the red-hot stove. Light a piece of paper and wave it to and fro over the hot stove. You will see the burnt particles rise, whirling and ascending sometimes as high as the ceiling. Why do they rise thus? They do so because they are carried by the air which, being heated by contact with the stove, becomes lighter and forms a rising current. These light fragments of burnt paper show us the upward flow of the air just as pieces of floating wood indicate the current of water. Thus air that is heated becomes light and rises.

"There we have the explanation of the draft in a fireplace. When the fire is lighted in the fireplace, the air contained in the chimney is warmed, becomes lighter, and rises. The hotter the air and the higher the column of heated air, the more powerfully it rushes upward. At the same time that the hot air rises, cold air, which is heavier, flows toward the fireplace, accelerates combustion, becomes warm in its turn, and joins the ascending column. In this way there is set up a continual current from the lower to the upper part of the chimney. To this incessant flow of air through the fireplace we give the name 'draft.'

"The prime requisites for a good draft can now easily be seen. First, the chimney must be entirely filled with hot air. If the channel is too large there is established at the top a descending current of cold air which mixes with the warm ascending current, slackens its course, and even makes it flow back into the room. Then the chimney smokes. A remedy for this is to make the chimney smaller at the top or else cap it with a sheet-iron pipe.

"Our chimneys are generally too large. Their faulty construction is necessitated by the method frequently employed for cleaning them. When a poor child of Savoy, all begrimed with soot, worms his way up the chimney by dint of much scraping of elbows and knees, in order to sweep the soot from the inner surface of the walls, the passage must be large enough to admit his body, even though the draft suffer in consequence. But if the sweeping is done in a more suitable manner with a small bundle of brushwood let down from above by a cord, there is no reason why the chimney should not be as narrow as may be necessary for a perfect draft.

"Often, too, the lower opening, the one from the chimney into the room, is too large. Then there enter the channel at the same time hot air from the central part where the fire is burning and cold air from the vacant lateral parts. This cold air necessarily lessens the draft by mixing with the hot air and lowering its temperature; or it can even blow the smoke back into the room.

"As far as possible only hot air should enter the chimney, all cold air sucked in by the draft being made to traverse the mass of burning fuel before passing into the ascending flue. To this end, in properly constructed fireplaces, the inner opening is narrowed by making the enclosing walls of the fireplace run obliquely inward so that most of the air sucked in by the draft passes through the burning fuel and becomes warm.

"The slanting walls serve still another useful purpose: they send back into the room a part of the heat that would not otherwise be reflected. To increase their efficacy in this respect they are lined with glazed tiles, which by their polish reflect a great part of the heat.

"Finally, the top of the chimney should be equipped in such a manner as to keep out any gust of wind that might else go whistling down toward the fireplace and thus drive back the smoke. To this end the chimney is capped either with a chimney-pot, which offers obstruction to the inflow of outer air, or with a sheet-iron hood that turns with the wind and always points its opening to the leeward.

"On account of the great volume of air continually passing through its wide mouth, a fireplace gives to a room excellent ventilation, a condition indispensable to health in our close dwellings; but its utilization of heat is sadly defective, so far as warming is concerned, because the air heated by passing through the fire is discharged into the outer atmosphere with no benefit to any one.

"It is just the opposite with stoves: they warm well, but they renew the air of a room very imperfectly. They warm well because the whole of their heated surface, that of the sheet-iron pipe as well as that of the stove itself, is in contact with the air of the room. Cast-iron stoves furnish quick and intense heat, but cool off as quickly if the fire dies down. Terracotta stoves, whether glazed or not, heat more slowly, but their action is more continuous, more gentle, more even; they retain their warmth a long time after the fire is out.

67

The Swedes and the Russians, in their rigorous climate, use enormous brick stoves occupying an entire wall of the room. The smoke and other products of combustion, before escaping out of doors, circulate through this mass of masonry by numerous channels. A fire is lighted in the morning and left to burn for several hours; then, when the wood is all converted into glowing coals, every outlet is closed, and that suffices to maintain a gentle heat in the room until night, provided only the glacial outside air be not admitted. But this mild and equable temperature is secured only at the sacrifice of the purity of the atmosphere, which cannot be renewed in the tightly closed room.

"Our stoves have the same fault: they do not renew the atmosphere of a room well because they consume for the same amount of fuel much less air than a fireplace. In a stove, in fact, all the air that enters is used up in burning the fuel; in an open fire, on the contrary, much air is drawn in that does not pass through the burning fuel, but escapes outside without having taken part in the act of combustion.

"Besides the disadvantage of furnishing poor ventilation, the cast-iron stove has still another defect. The intense heat that it throws out dries the air to such an extent as to make it unpleasant to breathe. The great thirst one feels near a very hot stove has no other cause. This dryness can be remedied by placing on the stove a vessel full of water, which in evaporating gives suitable humidity to the air. Finally, the various kinds of dust floating in the air burn on coming in contact with the red-hot stove, and give rise to disagreeable emanations. In short, if the stove is the best heating apparatus in respect to easy installation, economy of fuel, and utilization of heat, it is one of the most faulty from a hygienic standpoint, especially in a small room filled with many people."

Chapter Twenty-Five - Lighting

"THE crudest method of lighting I have ever seen is to be found in certain out-of-the-way corners of our mountain districts. Under the mantel of an immense fireplace there stands, on one side, a box of salt, kept dry by the heat of the fire, while on the other projects a large flat stone on which in the evening, with true economy, resinous splinters from the trunk of a fir-tree are burnt; and that is the only light, the only lamp, in the dwelling. By the light of this red and smoky flame the housewife prepares the soup and pours it into the porringers for the family returned from their labor in the fields, while the grandmother, silent in her big arm-chair, plies the spindle with her lean fingers."

"A common oil-lamp," said Claire, "would be far preferable to that little fire of resinous wood."

"I do not deny it," replied Uncle Paul; "but oil costs money, and a few little splinters of fir cost nothing."

"But that pitch-pine torch only gives light right round the fireplace. For going at night from one room to the next some other kind of light is needed."

"They have the iron lamp, in which a cotton wick inserted in a sort of socket burns grudgingly a few drops of nut oil."

"That must give a very poor light — not so good as a candle, even the kind that makes such a horrid smell."

"What are those bad-smelling candles made of, Uncle Paul?" asked Emile.

"They are made of either mutton or beef fat, which we know by the name of tallow."

"There are others, of a beautiful white, that are not greasy to the touch and have hardly any smell."

"Those are tapers, or wax-candles, as they are called, and they are far superior to common candles, although made from the same material, tallow. It is true that the tallow destined for tapers undergoes thorough purification, which removes its disagreeable odor and its oil. The first step in this process is the boiling of the tallow in water to which lime has been added. Next there is called into service a powerful drug, oil of vitriol, which takes away the lime as soon as it has acted sufficiently. Finally, the material under treatment is subjected to strong pressure which squeezes out and gets rid of the oily ingredient. After passing through these various stages the tallow is no longer what it was. It has no smell, or hardly any; its consistency is firm, its color a perfect white. In this new state, which is very different from the first, it is no longer called tallow but stearin.

"The notable improvement I have just described, which was destined to give us the odorless taper with its white flame in place of the foul-smelling, smoky tallow-candle, dates from the year 1831.

"Tapers, or wax-candles, are made in molds, which are of metal and open at their base into a common reservoir, the use of which I will explain presently. Through each mold, running from end to end, passes a wick, which is fastened at the bottom by a small wooden pin, and at the top by a knot that rests on the little central opening of a conical cap. After the molds have been placed in position, with their pointed ends downward and their bases upward, the melted stearin is poured into the reservoir, whence it runs into the several molds. Nothing more remains to be done except to bleach and polish the candles. They are bleached by exposure to the sunlight for some time, and polished by rubbing with a piece of cloth.

"Oil, which I will speak of in detail later, is also used for lighting. There are several inferior kinds commanding so low a price as to permit of their use. Those in most common demand for lighting are rape-seed oil, colza oil, and nut oil. They are burned by the use of a woven cotton wick which soaks up the liquid at its lower end and brings it drop by drop to the flame.

"If the candle is allowed to burn fully exposed to the air, with no protecting glass chimney to regulate and increase the draft, the flame is dim and smoky, a part of the oil, decomposed by the heat, being lost in smoke or collecting in the form of snuff on the incandescent part of the wick. To bum this smoke, this carbon, and thus obtain a brighter light, a good draft must be generated, as in fireplaces and stoves, so as to utilize to the full the fuel that feeds the flame. This is attained by means of a glass chimney enclosing and surmounting the flame, allowing cold air to enter at the bottom and hot air to escape at the top. Only on this condition do lamps, of whatever sort, give a good light."

"I thought," said Marie, "that lamp-chimneys were only to protect the flame from puffs of air."

"Chimneys do indeed protect the flame from puffs of air that might make it flicker and go out, but they also fill another office no less useful: they generate a draft without which combustion would be imperfect and the light dim. If you remove the chimney you will notice that the flame immediately becomes smoky and dim; put the chimney on again, and the flame will instantly recover its vigor and clearness. It is the same with the lamp as with the stove: without a sufficiently long pipe to create a draft, or, in other words, to draw in air to the very heart of the fuel, the stove would burn badly and give out little heat; and without the glass chimney bringing to the wick a continual stream of fresh air the lamp would make but poor use of its oil supply and would give only a faint light.

"The chimney is especially indispensable with lamps that burn kerosene, a liquid very much used at the present time for lighting. The lamp hanging from the ceiling of this room to give us light in the evening is filled with kerosene. You must have noticed the great difference it makes with the lighted wick whether the chimney is on or off."

"That difference has always struck me," Marie replied. "Before the chimney is on the flame is red, very smoky, and gives hardly any light; but as soon as the chimney is put on, the smoke disappears and the flame turns a very bright white. Without its chimney the lamp would not give much light and would fill the room with black smoke."

"In kerosene lamps the air necessary for combustion comes through a number of little holes in the metal disk that supports the chimney. If these holes become obstructed with the carbonized refuse of the wick, the draft works badly and the lamp smokes. Then sometimes, ignorant of this detail, you are lost in conjecture as to the cause of the wretched light you are getting. The wick is badly trimmed, you say to yourself, it is too long or too short, there is not enough oil, or there is too much. You try this and that remedy, all to no purpose; the lamp still smokes. The right remedy is nevertheless very simple: it is only to take off the metal disk through which the wick runs, and with a brush thoroughly to clean the little holes with which it is riddled. These openings clean, the air once more comes freely to the flame, a draft is created without any obstacle, and the lamp burns as brightly as ever."

Chapter Twenty-Six - Kerosene Oil

"WHERE does kerosene come from?" asked Claire. "Is it a kind of oil something like olive oil and nut oil?"

"Notwithstanding the name of oil generally given to it," her uncle replied, "kerosene has nothing in common with real oils. These are extracted from certain kinds of fruit, the olive for example, or from certain seeds like nuts and the seeds of flax and rape. Kerosene has no such origin; it is found ready-made in the bowels of the earth where certain stones are impregnated with it and let it ooze out drop by drop. The word petroleum (another name for it) alludes to this mineral origin, for it signifies stone-oil."

"Stone-oil! "exclaimed Jules. "I have never seen stones that let anything like oil ooze out of them."

"Nevertheless there are some, and in a good many countries, too. I have already told you the history of pit-coal. [1] You know that it comes from pre-historic vegetation buried to a great depth by the cataclysms which earth and sea have undergone. This coal, when heated without exposure to the air, gives the illuminating gas that burns with a white flame superior in bright-ness to the light of our best lamps; it also yields coal-tar, which contains in large quantities certain inflammable liquids, mineral oils closely resembling petroleum. Layers of vegetable matter buried by convulsions of nature in the bowels of the earth, and converted into something like coal, have managed to undergo in one way or another the kind of decomposition accomplished in our gas-works. Thence have come natural tar, bitumen, black pitch, which the miner encounters in his excavations; thence also comes the inflammable gas which in various countries escapes from fissures in the ground and serves as an inexhaustible supply of fuel; and thence too we derive the liquid fuel designated by the name of petroleum. So heat from coal and light from petroleum are both a heritage coming down to man through centuries from the ancient vegetation of the world."

"You speak," said Marie, "of the supply of fuel in decomposed vegetable matter buried in the interior of the earth. Does this supply furnish light and heat?"

"Most assuredly it does. There issues from it an inflammable gas, a gas that burns with an intense heat. Such supplies of natural gas are not rare in oil-fields, as for instance around Lake Ontario in North America. The inflamma-ble gas escapes through crevices in the rock, in the soil, and even in the bed of the lake itself. It catches fire at the approach of any lighted object and, where the volume of accumulated gas is large enough, sends forth long, bril-liant flames that even heavy rains sometimes fail to extinguish. The spectacle is wonderful beyond description when these jets of fire burst forth from the very bosom of the lake and dart hither and thither over the liquid expanse. In winter the effect is still more remarkable. Then around each gas-jet an ice chimney several decimeters long forms and takes the shape of a huge crystal candelabrum, at the top of which blazes the flame. The people in the neighborhood make use of this reservoir of natural gas: it is conducted to their dwellings through pipes, and there it is burned in the fireplace for cooking or in gas-burners for lighting. One of these burners gives as much light as four or five candles combined."

"Those jets of flame must be very odd," remarked Emile, "especially when they come out of the water; but Lake Ontario is a long way off, and I shall never see them."

"Without going so far as Lake Ontario," his uncle assured him, "you can see flames bursting out of the water in the nearest ditch. Choose one abounding in black mud made of decayed leaves, and stir that mud with a stick. Big bubbles of gas will rise and show themselves, bladder-like, on the surface of the water. Well, if you hold a piece of lighted paper near one of these bladders, the gas will catch fire, producing a slight explosion and giving a very feeble light which would be invisible in the full glare of the sun, but can easily be seen at nightfall or even in the shade. This inflammable gas comes from vegetable matter decaying under the water, just as the inflammable liquid petroleum has its origin in vegetable refuse buried in the depths of the earth.

"The same gas also escapes from coal, as is only natural, since coal is formed from the remains of ancient vegetation in the same way that black ditch mud is made principally of decayed leaves. Sometimes the subterranean excavations pushed to great depths for extracting the coal become filled with this inflammable gas. If a workman carelessly brings his lantern near this formidable substance — which gives no warning of its existence, as it is invisible and has no odor — a terrible explosion takes place, the mountain is shaken to its foundations, the scaffolding of the galleries falls down, and hundreds of people perish at such a depth, alas, that no succor is possible. Miners call this dangerous coal-mine gas "fire-damp."

"Around the Caspian Sea these fountains of fire, as we may call them, are numerous. All you have to do is to stir the earth a little way down and hold a light near the place, when instantly flames will spring up that will burn indefinitely. One of these fountains, that at Baku, is the object of superstitious reverence on the part of a religious sect, that of the Guebers or fire-worshipers. A magnificent temple encloses it. The inflammable gas gushes out through the cracks in the walls, from the summit of the vaulted roof, from the tops of the columns surrounding the edifice, from the very ground of the sanctuary, and from the entrance door. If a torch is applied, all these jets of gas catch fire in the twinkling of an eye and the temple is enveloped in a splendid curtain of flame."

"That would be better worth seeing, "said Claire, "than the inflammable bladders on the surface of a muddy ditch. But unfortunately, as Emile has already remarked, it is rather too far away."

"I hope I have said enough," Uncle Paul resumed, "to show you that the presence of a combustible liquid in the bowels of the earth has nothing in it that cannot be easily explained. Coal, inflammable gases, petroleum, all three have a common origin in the ancient vegetation of the globe buried and decomposed underground.

"Now let us talk about kerosene. North America furnishes most of it. Excavations similar to our wells are carried to a greater or less depth, and through the walls of these excavations the oil oozes, collecting little by little at the bottom as water does in our ordinary wells.

"Kerosene has an oily appearance, but is easily distinguished from oil by this peculiarity: whereas oil makes a translucent blot on paper and does not go away, kerosene makes a blot similar in appearance but disappears with heat without leaving any trace. That is because kerosene evaporates, while oil remains. Furthermore, the smell of kerosene is strong, penetrating, and somewhat like that of the tar that comes from our gas-works.

"The great inflammability of kerosene is a source of serious danger, which it is important to understand in order to exercise the prudence called for in handling this liquid. If you spill ordinary oil on the ground and hold a piece of lighted paper near it, you will not succeed in making it catch fire. Do the same with kerosene and it will take fire more or less quickly according to its quality. If it catches fire instantly it is a very dangerous liquid and should be used as little as possible, if one does not wish to be exposed to the risk of serious accidents. But if it takes fire with some difficulty, it can safely be used in our lamps. The best kerosene is the one that is slowest in catching fire."

"I should have thought, on the contrary," said Marie, "that the best would be the one that catches fire the easiest."

"It seems so to you because you overlook the danger attending a too high degree of inflammability. Suppose you are carrying a lamp filled with ordinary oil; you stumble and fall, and the oil, together with the lighted wick, is dashed against your clothing. What is the result? Nothing very serious. The oil spilt, being unable to catch fire even close to the wick, will soil your clothes, it is true, but at least you will not be burned. What a frightful risk you run, on the contrary, if the lamp is filled with kerosene, which takes fire so easily! Your face and hands are splashed with the terrible liquid, your clothes are soaked with it, and instantly you are all on fire and in imminent danger of being burned alive."

"Oh, goodness gracious! if that should really happen to one of us!"

"If that should happen to one of you either with kerosene or with any other inflammable liquid, such as alcohol or ether, the first thing to do would be to keep your presence of mind and not run about distractedly, this way and that, frightened out of your wits: for with your clothes flying in the wind thus created you would only make bad worse by fanning the flames. You should snatch up the first thing you can lay hands on, carpet, table-cloth, shawl, or cloak, and wrap yourself snugly in it so as to stifle the fire; you should wind it around your body very tightly and roll on the floor while waiting for some one to come to your aid."

73

"Who can flatter himself," said Claire, "that at such a time he will preserve his presence of mind?"

"You must do your best. Life may depend on it."

"Would it not be well," asked Marie, "to have nothing whatever to do with this dangerous liquid?"

"No, indeed, because we have nothing better for lighting purposes, and with a little prudence all danger is avoided. In the first place, we should use only kerosene that catches fire with difficulty when tested by pouring a little on the ground and applying a lighted match. Also, the supply should be kept in a tin can and not in a glass bottle, which might get broken; and when a lamp is filled it should be done at a distance from the fire. Finally, it is advisable to use this liquid as little as possible for hand-lamps, lamps that are carried from place to place as we carry a candle; it should be reserved for stationary lamps, those fastened to the wall or hanging from the ceiling and not likely to be touched after they are once lighted. In that way the inflammable liquid is not in danger of being spilt over us."

"These precautions make me feel safer," Marie rejoined; "but I know enough now to see that with kerosene prudence must never be forgotten."

[1] See "Field, Forest and Farm."

Chapter Twenty-Seven - Glass

"ONCE upon a time, so the story goes, some sailors overtaken by bad weather landed on a desert shore and lighted a big fire to dry and warm themselves and pass the night. There being no wood on this sandy coast, they made their fire of dry seaweeds and grass. The furious wind caused their fire to burn fiercely, and, behold, the next morning the sailors were greatly astonished to find in the midst of the ashes sundry lumps of a substance as hard as stone but as transparent as ice. If they had seen this substance at the water's edge, and in winter, they would certainly have taken it for ice, but, raking it out from under the ashes, they could not escape the conclusion that it was something else."

"And what was it?" asked Claire.

"It was glass, that precious substance that to-day gives us our window-panes, which enable us to keep our houses warm without excluding the daylight. The sailors scrutinized closely the deposit left after their fire had burned out, and they perceived that the great heat had caused a part of the ashes to fuse with the sand of the soil and thus produce the transparent substance in question. It was thus that glass-making was discovered."

"Was that long ago?" Jules inquired.

"This discovery, one of the most important ever made, dates so far back that only a very vague record of it has been preserved, and this record is probably a mixture of fact and fable. But whether it be fact or fable, the story teaches us at least one thing: sand melted with ashes produces glass. Now, what can the substance be that imparts to ashes the property of thus transforming the sand? What is there in ashes of such a potent nature as to bring about this wonderful change?"

"There is soda," Marie suggested, "that same soda that turns oil or tallow into soap."

"It was in fact the soda from the marine plants burned by the sailors that had brought about the melting of the sand and the formation of lumps of glass. Potash, which closely resembles soda in all its properties, acts in the same way when it is heated to a high temperature with sand. In both cases the result of the fusion is glass, more or less colored, finer or coarser, according to the purity of the materials used. The fine and perfectly colorless glass of our goblets, decanters, and flasks is obtained from potash and very white sand; window-glass, which is very slightly green, at least on the edge, is made of soda and pure sand; common bottle-glass, dark green in color, or nearly black, is made of very impure sand and ordinary ashes.

"The manufacture of window-glass is a very curious operation. In a furnace heated to a very high temperature are large earthen pots or crucibles filled with a mixture of soda and sand. When these two substances are thoroughly melted together the result is a mass of glass, red-hot and running like water. Each crucible is removed by a workman and his assistant, standing on a platform in front of an opening through which the crucible is withdrawn. This workman is called the blower."

"Why *blower?*" asked Jules. "Does he blow?"

"Indeed he does, and vigorously, as you will see. His tool is an iron rod or tube with one end cased in wood to enable him to handle the metal implement without burning himself. The assistant heats the other end by passing it through the opening in the furnace, and then plunges it into the crucible. In this way he gathers up a certain amount of pastelike glass, which he molds into globular form by turning it around -again and again on a block of wet wood. That done, he again heats the glass at the furnace opening, softens it, and passes the rod to the workman, the glass-blower.

"The latter first blows gently into the tube, and the mass of glass becomes inflated into a bubble exactly as soap-suds would do at the end of a straw."

"I can make beautiful soap-bubbles by blowing with a straw," said Emile. "Does the workman do it like that?"

"Yes, just like that. He blows through his tube into the mass of glass which, flexible and soft as long as it remains red-hot, swells into a bladder. Then the tube is raised aloft and the workman blows the glass above his head. The bladder becomes flattened a little by its own weight, at the same time gaining in width. The blower lowers the tube again and swings it to and fro like the pendulum of a clock, every now and then resuming his blowing with greater force. By the action of its own weight, which lengthens it, and the blowing, which distends it, the mass of glass finally assumes the shape of a cylinder.

"The completed cylinder ends in a round cap which must be got rid of. To accomplish this the end of the cylinder is held near the opening of the furnace to soften it, after which the top of the cap is punctured with a pointed iron. By swinging the tube this puncture becomes enlarged and the cap disappears. The cylinder, hardened now although still very hot, is next placed on a wooden frame containing a number of grooves or glitters for receiving the cylinders. With a cold iron the workman touches the glass where it adheres to the tube, and by this simple contact a break occurs along the line thus suddenly chilled, leaving the cylinder entirely freed from the tube.

"Notice, children, the clever device adopted by the workman for detaching the glass from the tube without shattering it. He merely touches the very hot glass with a cold iron, and that suffices to produce a clean break all along the line touched. Glass possesses this curious property of not being able to withstand a sudden change in temperature without breaking. Chilled suddenly, it breaks; heated suddenly, again it breaks. That is a warning to you when you wash drinking-glasses or other glass objects. Beware of hot water if these objects are cold, and of cold water if they are hot; otherwise you run the risk of breaking them instantly. When the cold or the heat acts only along a predetermined line, it is on that line, suddenly chilled or heated, that the rupture occurs. That is how, without the slightest difficulty, the workman separates the glass cylinder from the iron tube to which it adheres.

"This done, the next thing is to remove the cap that still terminates one end of the cylinder. To do this the workman encircles this cap with a band of very hot glass, and then touches with a cold iron the line thus reheated. Instantly a circular rupture detaches the cap. Thus there is left on the frame a glass muff open at both ends. To split this muff the workman draws lengthwise, from one end to the other, a red-hot iron point, and then touches the hot line with a wet finger. A cracking follows, and the muff splits open. It is next taken to a furnace where, after being softened sufficiently, it is laid open and flattened out with an iron rule on a cast-iron plate. The final result is a large sheet of glass which the glazier will cut later with a diamond point into panes of any desired size."

"That is a very curious operation you have just told us about, Uncle Paul," said Claire. "For my part, I should never have suspected that a pane of glass, so perfectly flat as it is, was first a glass ball blown out like a soap-bubble."

"And how are bottles made?" asked Jules.

"For bottles glass is both blown and molded. The tube, laden by the assistant with the proper amount of melted glass, is passed to the blower, who gives to the vitreous mass the shape of an egg ending in a neck. The piece is then re-softened in the furnace and put into an iron mold. By energetic blowing the workman inflates the glass and makes it exactly fill the mold. This operation leaves the bottom of the bottle still flat, but by pressure with the point of a sheet-iron blade this bottom is driven up inside and shaped like a cone. A band of melted glass applied to the narrowed opening of the piece gives the neck of the bottle. The seal that some bottles bear, for example

76

where the word liter is inscribed, is made by attaching a small disk of glass while it is still soft, and stamping it with a mold made of iron suitably engraved."

Translator's Note. — Since the foregoing was written, machinery has largely displaced hand-work in glass-making.

Chapter Twenty-Eight - Iron

"OF all substances iron displays the greatest power of resistance; and it is this property that makes it the most useful of all our metals. All sorts of tools are made of it, and without it our industries would come to an immediate standstill."

"Yes, I see that iron is very useful," said Jules.

"But perhaps you do not see how useful it is. Reflect a moment and you will perceive that iron has had something to do with nearly everything in the house you live in. And at the very outset, in order that the house may be built, there is need of stone from the quarries, where picks and crowbars and chisels and hammers are in constant use, all these being, as you know, made of iron or steel. The beams and joists of the framework come from trees felled by the woodman's axe; and these timbers are squared and shaped, fitted and adjusted with the help of various carpenter's tools, all made of the metal we are considering. Hardly a single one of our articles of furniture would be practically possible without this metal: we need the saw for cutting the log into boards, the plane and the draw-knife for smoothing the surface, the auger and the bit for boring the holes that are to receive the wooden pegs which hold the parts together.

"Our daily sustenance entails the use of iron in an almost equal degree: it calls for the spade, the hoe, and the rake for working the garden which gives us our vegetables, and the plowshare for the heavy work of the field which furnishes us with bread.

"In the clothes we wear, too, we are hardly less dependent upon this indispensable substance. Of it are made the shears that clip the wool from the sheep's back, and of it also are made the carding and spinning and weaving-machines that convert the wool into cloth. Our most delicate fabrics, our ribbons and laces, require the aid of this metal in their manufacture; and, finally, is not the needle used in every stitch that is sewed, the needle so fine and sharp-pointed that nothing else could take its place?"

"It is plain enough," said Marie, "that iron is of the utmost importance to us, though we usually give it very little thought. It is so common that we make no account of it, in spite of the immense service it renders us."

"I should like to know how iron is obtained," Claire here interposed, "if Uncle Paul will tell us."

"Iron ore," he explained, "is a yellow or reddish stone of very unpromising appearance, with no resemblance to the metal so familiar to us. The furnace in which it is worked is a sort of high tower, swollen toward the base, tapering at the two ends, and measuring at least ten meters, sometimes twenty, in height. Through the upper door or mouth of the furnace is poured coal by the cart-load, with fragments of ore, and when once lighted the furnace burns uninterruptedly, day and night, until the masonry succumbs to the intensity of the heat. Workmen are continually piling on fuel and ore as fast as there is any subsidence in the burning mass, while other workmen at the base of the furnace watch the melting of the ore. Enormous blowing-machines inject a continuous stream of air into the lower part of the furnace by a great tube, through which this current passes, not as a gentle breeze, but as a veritable tornado, raging and howling with an uproar that is fairly deafening. If one takes a peep into the furnace from the opening by which the tube enters, a sort of white-hot inferno dazzles the eye. Here stones melt like butter, and iron, separating from the impurities mingled with it, falls in glowing drops into a reservoir or trough at the base of the furnace. When the trough is full a passage is opened by the removal of the clay stopper that closed it, and the liquid metal runs in a fiery stream into channels prepared for it in the ground.

"The metal thus obtained is impure iron and is known as cast-iron. It is run into molds to make stoves, grates, pots, and kettles, chimney-plates, water-pipes, and countless other objects. Although of great hardness, cast-iron is brittle: it breaks easily when sharply struck."

"One day," said Emile, "a stove-cover broke into three pieces just from falling on the floor."

"It is well to bear in mind," his uncle observed, "that all our cast-iron implements are more or less fragile, and that it takes but a blow or a fall to break them."

"Then," remarked Claire, "cast-iron won't do for making anything that must resist violent shocks — hammers, for instance."

"No, cast-iron is worthless for tools that are subject to rough handling; pure iron alone has the necessary resisting power. To purify cast-iron and convert it into wrought-iron, the workers heat it in still another furnace; and when it has become quite red and soft it is hammered with a block weighing some thousands of kilograms, and which is raised by machinery and then falls with all its weight. At each blow from this enormous hammer what is not iron escapes from the rest and runs off in a sweat of fire.

"After this hammering the mass is grasped between two cylinders, one above the other, turning in opposite directions. Dragged along by this powerful wringing-machine and flattened out by the irresistible pressure, it becomes in a very short time a uniform bar of iron. Shears next take hold of the bar and cut it up into pieces of equal length."

"What! Are there shears that can cut iron bars?" exclaimed Claire.

"Yes, my child; in those wonderful factories where human invention is brought to bear on the working of iron there are shears that without the least appearance of effort cut clean through a bar of iron with each snip, no matter if the bar be as big around as a man's leg. With our scissors we could not more easily cut a straw."

"But such shears cannot be operated by hand?"

"Nor are the two blades moved at the same time. While one of them remains at rest on a support, the other goes up and down as calmly and noiselessly as you please, cutting with each downward stroke the bar of iron offered it by a workman."

Chapter Twenty-Nine - Bust

"THE dull red that dims the luster of polished iron or steel is, as you know, rust, or, in learned language, oxide of iron; that is, iron mixed with oxygen. As it comes from the mine, iron takes this form of rust mingled with stone. What an unprepossessing appearance it then wears, this most useful of all the metals! It is an earthy crust, a reddish lump, a shapeless mass, in which the presence of any kind of metal whatever can be divined only after painstaking research. And then it is by no means enough to determine that this rusty matter contains a metal; it is still necessary to find some means of decomposing the ore and extracting the iron in its true metallic state. What labor and experiment has it not required to attain this end, one of the most difficult imaginable! How many fruitless attempts, how many laborious trials!

"Other metals, the greater number of them, likewise rust, the color of the rust varying with the metal. Iron turns a yellowish red; copper, green; lead and zinc, white. Take a knife and cut through a piece of lead. The cross-section shows a fine metallic luster, but before long it becomes tarnished and a sort of cloudy appearance is noticeable. This change begins as soon as there is contact with the air, and in course of time it extends, slowly indeed, but surely, until it penetrates to the very heart of the mass and ends by converting the lead into an earthy substance quite different in quality. And the greater number of metals undergo a similar deterioration under like conditions."

"Then that green stuff we see on old copper coins is rust?" asked Marie. "And the whitish coating on the water-pipe of the pump?"

"Yes, it is rust in each instance. But all metals are not equally subject to rust. Iron is one of those that rust most quickly; next come zinc and lead; in the third class are tin and copper; and in the fourth is silver, which remains free from rust with very little care; finally, gold is still more immune and never rusts.

"Gold coins and jewelry of the remotest antiquity come down to us as pure and brilliant as if made yesterday, despite a sojourn of long ages in a damp soil where other metals would have turned to shapeless rust. Since it has such power of resistance to destructive agencies, gold ought to be found, and in fact is found, always retaining its metallic properties, especially its luster. In the bosom of the rocks where it is disseminated — in its ore, as we say — it forms scales, veins, and sometimes big nuggets, which shine like jewels just from the goldsmith's hands. Our ear-rings and finger-rings, carefully kept in their casket, are not more brilliant than the particles of this precious metal found in the heart of a rock. Just as it occurs in its natural state it can be put to immediate use; all that is needed is to hammer it and shape it. Hence it is the first of metals to be discovered and used by man; yet owing to its extreme rarity it has never, in our part of the world, been used for common tools, but has ever remained the preeminently precious metal, reserved for jewelry and coinage."

Chapter Thirty - Tin-Plating

"IRON," said Uncle Paul, "is abundant and cheap; A furthermore, the hottest fire in our stoves and grates cannot melt it, and it is able to withstand rather rough usage. These qualities are highly important in cooking-utensils, which must resist the action of fire without risk of melting, and have every day to undergo bangs and falls. But unfortunately this metal rusts on the slightest provocation; contact with a few drops of water for any length of time suffices to cover it with ugly red spots which eat into it and finally pierce quite through. This rapid deterioration is prevented by tinning.

"A metal is tinned by being overlaid with a thin coating of tin, which resists rust. Now bear in mind this important point, which I have already briefly touched upon: rust develops only where air is present. Also, it is promoted by various substances, such as water, vinegar, and the juice of our vegetables and fruits. Nearly all the dishes we prepare for the table tend by their mere contact to rust metals capable of rusting, and especially iron. To prevent the formation of rust, therefore, what must we do? The answer is plain: keep our food and the air from coming in contact with any metal that will rust. If this contact never occurs, rust will never form, for there will be nothing to cause rust.

"Our obvious course, accordingly, is to coat the corruptible metal with one that will protect it, and this last must fulfil two conditions: it must not be liable to rust; or, at any rate, it must be a metal that rusts with difficulty, since otherwise one ill would only be exchanged for another; and it must also be a metal from which food will contract no injurious properties. This double requirement is met by very few metals. There are, first, gold and silver, both too expensive for common use; and finally there is tin. This metal is very slow to rust, and, furthermore, tin-rust, if it ever begins to form, does not form in any quantity and has, besides, no harmful properties. Tin, therefore, furnishes us the metallic coating we need for preserving our iron utensils from rust."

"Wouldn't it be much simpler," asked Claire, "to make these utensils wholly of tin in the first place and so get rid of iron altogether?"

"There is one serious objection to such a course: tin melts easily. A saucepan of this metal would not hold out for five minutes against the heat of a handful of glowing charcoal. What would become of your stew in" a cooking-utensil capable of melting like wax over the fire!"

"I see now that tin by itself would never do."

"Nor would it answer for another reason: it offers too little resistance, it bends under slight pressure, it is knocked out of shape with a blow. We must have two metals combined: iron to resist heat and stand rough usage, and tin to prevent rust. If, however, the utensil is not to go over the fire, it can in case of need be made of tin alone. Not very long ago, in the country, tableware for company use was of tin. Plates, platters, and soup tureens shone like silver on the shelves of the dresser, and were the pride of the housewife. Our measures for wine, oil, and vinegar are of tin. The use of this metal in preference to any other for utensils that are to come in contact with our food is due to its perfect harmlessness. Tin keeps its cleanness and polish and — a still more valuable property — communicates nothing injurious to the substances it touches.

"You are familiar with those wandering tinkers with sooty faces who, a kettle over the shoulder and a few old forks in one hand, go through the street, crying their trade in a shrill voice. In the open air, over a little charcoal fire, they restore rusty covers to their first brilliance, mend kettles and saucepans, and plate with tin utensils of iron and copper, to keep them from rusting. The operation is very simple: the piece to be tinned is first well scoured with fine sand, and then heated over the fire, and while it is still warm a little melted tin is rubbed over the surface with a wad of tow. The tin takes fast hold of the underlying metal and covers it with a thin layer which will not come off with rubbing. That is what is known as tin-plating."

"Then what we commonly called tin," said Marie, "is really iron covered over with tin?"

"Yes, it is tin-plated iron, and is made by plunging thin sheets of iron into melted tin. These tinned sheets, light and strong at the same time, polished and rust-proof, serve for the manufacture of numberless utensils. The outfit of our kitchens consists in great part of tinware."

"But it consists also of copperware." remarked Jules.

"Yes; but copper has very dangerous properties which call for the utmost caution on our part. Iron rust is harmless, I might even say healthful, in limited quantities. Little children deficient in bodily vigor are sometimes made to drink water impregnated with a small quantity of rust from a few old nails

81

in the bottom of the water-bottle. Nothing, then, is to be feared, so far as our health is concerned, from the rusting of iron; it is coated with tin, not as a safeguard against danger, but to give the metal cleanliness and greater durability.

"Copper-rust, on the contrary, is a violent poison. This rust, or verdigris, is all the more dangerous in that it develops with extreme ease when copper comes in contact with our articles of food, especially when the latter contain vinegar or fat. Have you ever noticed the greenish tinge imparted to the oil in lamps and to the candle-drippings on candlesticks? Well, this tinge comes from the copper that enters into the metal part of lamps and candlesticks, and is due to the verdigris dissolved in the oily matter. Our food, containing as it almost always does some slight proportion of fat or oil, contracts the same greenish tinge by remaining any length of time in contact with copper. Vinegar acquires it in a few moments. This green substance from copper, always bear in mind, is a terrible poison which cannot be too carefully shunned. The only safe course lies in constant watchfulness and in a scrupulous cleanliness that keeps all copper utensils always bright and free from the slightest indication of a green speck. For greater safety it is even preferable to use only tinned ware for culinary purposes. Proof against rust under its coating of tin, copper then ceases to be injurious, but only on condition that the tin coating is maintained intact, never uncovering a particle of the poisonous metal. As soon as the tin becomes dim and shows a little of the red underneath, the utensil should be retinned.

"Lead is no less dangerous than copper, but this metal enters very little into domestic use, unless it be for cleaning bottles. A handful of small shot shaken up in water serves excellently to remove by friction the impurities clouding the inside of the glass. This practice, however, is not free from one grave danger. Suppose a few particles of lead are retained in the bottom of the bottle and left there unobserved. Wine, vinegar, or whatever other liquid is afterward poured in, is likely to cause the lead to rust, and will thus contract properties highly injurious to health. Without any one's suspecting it there will lurk in the bottom of the bottle that is daily used a permanent source of poison. You see therefore what care should be exercised in order that not a particle of lead may be left behind after this metal has been used for cleaning purposes. Never forget that copper and lead are two poisonous metals, and that any carelessness in their domestic use may suffice to imperil our very lives."

Chapter Thirty-One - Pottery

"TRAVELLERS tell us of the strange methods M. adopted in preparing food by certain savage tribes among whom that precious culinary utensil, the earthen pot, is unknown. They tell us, for example, that the Eskimos of

Greenland boil their meat in a little skin bag, as I have already [1] related to you. The bag is not placed over the fire, but is filled with water and the food to be cooked, after which stones heated red-hot are dropped into it; and thus, laboriously and imperfectly, by repeated heating of the stones and dropping them into the bag, a dish of half-cooked meat mixed with ashes and soot is at last prepared.

"To such extremities we might be reduced if it were not for those little earthen pots, sold at a penny apiece, that rescue us from the dire straits familiar to the Eskimo. Let us now consider how this simple kitchen utensil is made.

"From the modest little porringer to the sumptuous porcelains adorned with rich paintings, every piece of pottery is made of potters' earth, or clay, which is found almost everywhere, but by no means of uniform quality. There are yellow clays, red clays, ash-colored clays, dark clays, and perfectly white clays. These last are free from all foreign matter; the others contain divers alien substances. All are easily kneaded with water, forming a sort of unctuous dough capable of taking any prescribed form. The coarsest clays serve for making bricks, drain-pipes, flower-pots, and so on; clays that lack purity but are still of fine texture are used for common pottery; and, finally, clays of extreme purity, of snowy whiteness, furnish us porcelain. This degree of purity is very rare in clay, being found in. France only in Haute-Vienne, around Limoges. Clay of inferior quality occurs in abundance in nearly all parts of the world.

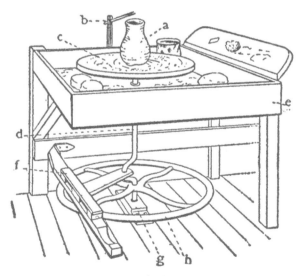

Potter's Wheel

a, partly molded clay; *b*, guiding measure; *c*, revolving wheel, screwed on shaft *d*, which is propelled by horizontally moving treadle-apparatus *f*, and steadied by fly-wheel *h*, pivoted on block *g*; *e*, box for holding balls of clay, water-vessel, sponge, tools, etc.

"In order to give to moistened clay quickly and easily a regular' form the potter makes use of the potter's wheel. As illustrated in the picture I here show you, under the potter's work-table is a horizontal wooden wheel which the operator sets in motion with his foot. The axle of this wheel carries at its upper end a small disk, in the center of which is placed the lump of clay that is to be shaped by the potter. The latter thrusts his thumb into the formless mass, which rotates with its supporting disk, and this action suffices to produce a symmetrical cavity because of the regularity of the motion imparted to the clay. As fast as the thumb enlarges the cavity the other fingers are applied to the outside to hold the mass in place, to give it the desired shape, and to preserve a uniform thickness of wall throughout. In a few moments the piece is fashioned, and we see the lump of clay hollowed out and made to stand up in the form of a bowl or jar having just the outline and thickness desired by the artisan. The application of the palm of the hand, slightly moistened, suffices to polish the surface. Finally, with tools designed for the purpose the piece is ornamented with moldings. For example, it is enough to touch the rotating object with an iron point to trace an engraved line around it.

"When the potter's wheel has done its part the vessel, still damp, is left in the air to dry, after which it is dipped into a bath of water and fine dust of lead ore. By the action of fire this dust will presently be incorporated with the surface clay on which it rests and will become a sort of glaze or varnish, without which the vessel would be permeable by liquids and would allow its contents gradually to ooze out and escape. To complete the whole the vessel is subjected to a high temperature in an oven, where the clay bakes and becomes hard stone, while the lead dust covering it melts and combines with the substance of the clay, spreading over the surface as a brilliant varnish having the color of honey. In this wise the more ordinary pottery is made, the pottery constantly used in our kitchens and so valuable for its ability to bear heat without breaking.

"In most cases the coating of lead has nothing to be said against it, for our articles of food do not, as a rule, produce any effect upon it. Vinegar alone is of a nature to dissolve it slowly when kept long in contact with it, especially if the vessel in question has not been properly baked. Hence it would be highly imprudent to use pots or jars having this lead glaze on the inside for keeping gherkins, capers, and other pickles preserved in vinegar. This latter might in course of time dissolve the metal contained in the glaze and thus contract poisonous properties, so that in seasoning a dish with a handful of capers one might run the very serious risk of lead-poisoning. Pickles of that sort should be kept in glass jars or in common earthenware not glazed on the inside.

"Crockery is made of clay of fine quality. Its glaze, which is of a beautiful milky whiteness, is prepared from tin, a harmless metal. Thus our food, even when containing vinegar, never contracts injurious properties from contact with this glaze.

"I will say as much for the glaze of porcelain, which contains no metal in its composition, but results from the melting of the surface clay itself in the heat of the oven, the clay being of extreme purity and whiteness. Hence we here have no fear of any lead varnish, which is recognizable from its honey-yellow hue and is employed only for common pottery. But, as I said before, this latter kind of glaze is to be feared only in case of prolonged contact with vinegar."

[1] See "Our Humble Helpers."

Chapter Thirty-Two - Coffee

"THE plant that produces coffee is called the coffee-tree. This is in reality little more than a shrub, bearing some resemblance to a small pear tree in its rounded top and bushy branches. Its leaves are oval and shiny; its blossoms, which resemble those of the jasmine and have a sweet smell, are grouped in small bunches in the axils of the leaves. These blossoms are succeeded by berries, first red and then black, having the appearance of our cherries, but with very short stems and crowded close together. Their pulp is insipid and sweetish, enclosing two hard seeds, round on one side, flat on the other, and united by their flat sides. These seeds are the so-called coffee-beans which we use after roasting them in a sheet-iron cylinder revolving over the fire. In color they are half-way between white and green, but turn to a chestnut hue in roasting.

Fruiting Branch of Coffee Plant

a, flower; *b*, section of berry, showing enclosed nutlets.

"The coffee-tree can thrive only in very warm countries. It is indigenous to Abyssinia, where it grows abundantly, especially in the province of Kaffa, from which it seems to have taken its name." In the fifteenth century the coffee-tree was introduced from Abyssinia into Arabia, and it is there that the plant has found a climate most favorable for the development of its peculiar properties. Indeed, the most highly prized coffee comes to us from the southern provinces of Arabia, especially from the neighborhood of Mocha."

"Then," said Marie, "when we speak of a coffee of superior quality as Mocha, we give it the name of the town that furnishes the best."

"Precisely. Look at the map and you will find Mocha very near the southern end of Arabia, at the entrance to the Red Sea. It is in this corner of the earth, under a burning sun, that the most highly esteemed coffee ripens.

"The Dutch were the first Europeans to turn their attention to coffee; they introduced it into their East Indian colonies, notably Batavia, whence a few young trees were sent to Amsterdam to be cultivated in hothouses, as the climate of Holland would by no means ensure the thriving of this warmth-loving shrub in the open air.

"One of these young trees was given to the Botanical Garden of Paris, where care was taken to multiply it under glass, and one of the plants thus obtained was given to Déclieux, who started for one of our colonies, Martinique, with his little coffee-tree rooted in a pot. Never, perhaps, has the prosperity of a country had so humble a beginning: this feeble coffee plant, which a sunbeam might have dried up on the way, was to be for Martinique and the other Antilles the source of incalculable riches.

"During the journey, which was prolonged and made difficult by contrary winds, fresh water ran short and the crew was put on the most meager rations. Déclieux, like the others, had only one glass of water a day, just enough to keep him from dying of thirst. The young coffee-tree, however, needed frequent watering in that extremely hot climate. But how water it when thirst devours you and every drop of water is counted? Déclieux did not hesitate to devote his own scanty allowance to the needs of his charge, one day giving it the whole glass, and the next going shares with it, preferring to suffer the most painful of privations in order to reach his destination with the young coffee-tree in good condition. And this satisfaction was not denied him. Today Martinique, Guadeloupe, Santo Domingo, and most of the other Antilles are covered with rich coffee plantations, all owing their origin to the feeble plant imported by Déclieux."

"That plucky traveler is a man to be admired," declared Jules, "and every time I drink coffee I shall think of him and what he did."

"Nothing in our part of the world can compare in beauty with an orchard of coffee-trees bearing simultaneously as they do, throughout most of the year, leaves of a lustrous green, white blossoms, and red berries, vegetation in those sun-favored regions knowing scarcely a moment's repose. Over the perfumed tops of the trees hover butterflies whose wings, as large as both hands, astonish one with the magnificence of their coloring. In the forks of the topmost branches the humming-bird, a living jewel, builds its nest of cotton, half the size of an apricot. On the bark of old tree-trunks great beetles shine with more radiant splendor than the precious metals. In an atmosphere laden with sweet odors negroes carrying baskets on their arms go through the plantations from one coffee-tree to another, carefully gathering the ripe berries one by one so as not to disturb those that are still green. Scarcely is this harvest gathered when other berries redden, and then still others, while fresh buds form and new blossoms open.

"The coffee-berries, or cherries, as they are called, are passed through a kind of mill which crushes and removes the pulp without touching the seeds. Then these are exposed to the sun. Every evening, to protect them from the dew, they are piled in a heap and covered with large leaves, to be spread out again the next morning. When the drying process is finished they are winnowed, the spoiled seeds rejected, and the harvest is ready for export."

"After that," said Claire, "the coffee only has to be roasted and ground and it is ready for use. Does any one know who was the first to use it?"

"According to a tradition current in the East, the use of coffee goes back to a certain pious dervish who, wishing to prolong his meditations through the night, invoked Mohammed and prayed to be delivered from the need of sleep."

"A pious dervish, did you say?" Emile interposed. "I don't know what a dervish is."

"It is the name given in Oriental religions to certain men who renounce the world and devote themselves to prayer and contemplation."

"And Mohammed?"

"Mohammed is a celebrated character who about twelve centuries ago founded in Arabia a religion that has now spread over a great part of the world, especially in Asia and Africa. This religion is called Mohammedanism or Islamism, and Mohammed is often designated by the title of Prophet.

"To return to the dervish who wished not to sleep that he might have so much the more time for prayer and meditation, he addressed his petition to Mohammed, and the Prophet appeared to him in a dream, advising him to go in quest of a certain shepherd. This man told the dervish that his goats remained awake all night, leaping and capering like fools, after having browsed on the berries of a shrub that he pointed out to him. It was a coffee-tree, covered with its red fruit. The dervish hastened to try on himself the singular virtue of these berries. That very evening he drank a strong infusion of them, and, lo and behold! sleep did not once come to interrupt his pious exercises all night long.

"Rejoiced at procuring wakefulness whenever he desired it, he shared his discovery with other dervishes, who in their turn became addicted to the sleep-banishing drink. The example of these holy persons was followed by doctors of law. But before long it was discovered that there were stimulating properties in this infusion used for dispelling sleep, whereupon it began to find favor with those who had no desire to be kept awake, until finally the bean chanced upon by the goats came into general use throughout the East.

"I advise you not to yield a blind belief to this popular tradition, for in reality it is not known by whom or in what circumstances the properties of coffee were first discovered. One point only remains incontestable, and the story of the dervish brings it out well: it is the property coffee possesses of keeping the mind active and driving away sleep."

"Coffee, then, really prevents one's sleeping?" asked Marie.

"Yes, but not every one feels this singular influence in the same degree. There are some not affected by it at all, and others, of a delicate and nervous temperament, who cannot close their eyes all night long if they happen to take coffee in the evening."

"And how about taking it in the daytime?"

"In that case the same objection does not present itself; there is even an advantage in having the mind fully active, especially if one is engaged in mental work. But for the most part coffee is a simple stimulant that favors digestion and excites new vigor. Long habit makes it for many a drink of prime necessity.

"Prepared from the green berry just as it comes from the country that produces it, the infusion of coffee is a greenish liquid, odorless and tart, which acts powerfully on the nerves."

"Is that the way," asked Claire, "that the dervish, taught by the capering of the goats, took his first cup of coffee?"

"Probably. Nothing but the ardent desire to combat sleep could have induced him to continue the dose, for the drink prepared in this way is very far from being palatable. The qualities that make us desire coffee, especially its fragrant aroma, are developed only by roasting; hence this operation should be performed with a certain degree of care. If insufficiently roasted, the coffee-beans remain green inside; then they are hard to grind in the mill and give a greenish yellow infusion with no aroma. Roasted too much, they are reduced to charcoal on the outside; then the infusion is very dark, bitter-tasting, and without aroma. Coffee is roasted to a nicety when it gives out an agreeable odor and has taken on a dark chestnut color.

"Coffee should be ground fine so as to yield its soluble ingredients readily to the water; and, finally, the infusion should never be boiled, because then the aromatic principle is dissipated, being carried away by the steam. Coffee allowed to boil would soon be nothing but a bitter liquid bereft of the qualities that give it its value. The best temperature is the one that approaches the boiling point but never quite reaches it.

"The high price of coffee has given rise to many attempts to substitute cheaper home-grown products for the precious berry. It has become customary to roast chicory roots, chick-peas, and acorns, to mix with the ground coffee. The only resemblance to coffee possessed by these various substances lies in the burnt odor, the chestnut color, and the bitter taste, with none of coffee's efficacious properties. Allured by the hope of gain, the merchant may exalt in high-sounding terms of praise the virtues of these cheap substitutes, but you may be sure he never has any of them served at his own table."

"They say," Marie remarked, "that coffee sold already ground is sometimes mixed with one of those worthless powders you speak of."

"That is only too true. This fraud can be avoided by buying the coffee-beans either already roasted or green, and in the latter case roasting them oneself."

Chapter Thirty-Three - Sugar

"COFFEE calls for sugar," resumed Uncle Paul the next day. "Who can tell me what sugar is made of?"

All remained silent until Emile rather hesitatingly ventured to say:

"I have heard, Uncle, that it is made of dead people's bones."

"And who told you that, you simple child?"

"Oh, a friend of mine," replied Emile, in some confusion at this strange notion, the falseness of which he now began to suspect without being able to explain it.

"Your friend," said his uncle, "was making game of your credulity when he told you any such ridiculous story as that. Sugar has no lugubrious origin of that sort, although there is a grain of truth in what your friend said. To purify sugar and make it as white as snow, use is made of animals' bones after they have been burnt to charcoal, as I will explain to you presently. But these bones, as soon as they have played their part, are thrown away and not the slightest trace of them is left in sugar as it comes to us. It is probably this use of bones in the manufacture of sugar that has given rise to the singular idea you repeat after your friend.

"Then there isn't one of you that knows where sugar comes from. But you do know at least many kinds of fruit that have a very sugary taste, such as melons, for example; grapes, figs, and pears."

"Melons are so sweet," put in Claire, "one would think they were preserved in sugar. Very ripe pears, too, are just as sweet; and so are grapes and figs."

"If these various kinds of fruit have the sweet, sugary taste to such a high degree, it proves that they contain sugar in their juice and in their pulp."

"And yet we don't sweeten them; we eat them just as they are."

"We do not sweeten them ourselves, it is true, but somebody does it for us; and that somebody is the plant, the tree that bears them. With a few poor materials which the roots derive from the soil, and with the drainage from the dunghill, the plant, an inimitable cook, concocts a certain amount of sugar and stores it up in the fruit for our delectation. Emile was inclined to believe, rather against his will, that sugar was made from dead people's bones; but here we have quite a different explanation of the matter: I tell him that the toothsome dainty really comes from certain filthy substances mixed with the soil under the form of manure. Before becoming the exquisite seasoning of the peach, fig, and melon, the sugary matter was nothing but foul refuse. This ignoble origin is not peculiar to sugar: everything offered us by vegetation is derived from a similar source — everything, even to the sumptuous coloring and the sweet perfume of flowers. To effect this marvelous transformation man's skill would be powerless; the plant alone is capable of such a miracle. Out of a few materials which the earth, water, and air supply, it makes an infinite variety of substances having every sort of flavor and smell — in fact, all imaginable qualities. For that reason I have called it the inimitable cook.

Man, then, does not really manufacture the sugar; it is the plant, the plant alone, that produces it, and man's work is limited to gathering it where he finds it ready-made and separating it from the various substances accompanying it.

"I have already mentioned several kinds of fruit as containing sugar, the melon in particular. Often other parts of plants contain it too. Chew a stalk of wheat when it is still green, or of reed-cane, or the first blade of young grass. You will find they have a slightly sugary taste. There is not a blade of grass in the meadow but has its stalk preserved in sugar. In other plants it is the root that becomes the storehouse of saccharine matter. Couch-grass, the commonest weed in our fields, has a very sweet root. The enormous root of the beet is sweeter still, being a veritable candy shop, so much sugar does it contain. You see how widely dispersed sugar is throughout the vegetable kingdom, although few plants lend themselves to the industrial extraction of this precious substance, because they contain so little of it. Two plants only, incomparably richer than the rest, furnish nearly all the sugar consumed the world over; and they are the sugar-cane and the beet-root.

Sugar-Cane

a, part of the inflorescence;
b, a spikelet.

"The sugar-cane is a large reed two or three meters high, with smooth, shiny stalks having a sweet, juicy marrow. The plant came originally from India and is now cultivated in all the warm countries of Africa and America. To get the sugar, the stalks are cut when ripe, stripped of their leaves, made into bundles, and then crushed, in a kind of mill, between two cylinders turning in opposite directions a short distance apart. The juice thus obtained is sometimes called cane-honey, which shows you how sweet it is. It is put into large kettles and boiled down to the consistency of syrup. In the course of the process a little lime is added to clarify the syrup and separate the impurities from it. When the evaporation has proceeded far enough the liquid, still boiling, is poured into cone-shaped earthen molds; that is to say, molds having the shape of a sugar-loaf. These molds, turned point downward, have at this end a small hole that is kept stopped up with a straw plug. As soon as they are full of syrup they are left to cool slowly. Little by little the syrup crystallizes and becomes a compact mass, after which the straw plug is removed and the small amount of liquid that has not hardened escapes through the hole at the point, drop by drop. This first operation gives raw sugar, commonly called brown or moist sugar. Its color is not yet pure white, and there is something disagreeable about the taste. To make it per-

90

fectly white and to free it from certain ingredients that mar the perfect quality of its flavor, it undergoes a purifying process in factories called refineries.

"In France sugar is obtained from the beet-root, an enormous root with white flesh, cultivated in vast fields for the manufacture of sugar in several of our northern departments."

"The beets I usually see in the field," said Marie, "have red flesh. Do they also contain sugar?"

"Yes, they contain some, but less than the white beets. Besides, their red coloring would add to the difficulty of obtaining perfectly white sugar. So the white beets are preferred. The roots are carefully washed and then reduced to pulp under large graters worked by machinery. Finally this pulp is placed in woolen sacks and subjected to pressure. The juice thus extracted is treated like that of the cane and yields a similar raw or brown sugar, which must be refined in order to attain perfection.

"The process of refining is based on a certain property of charcoal which you must learn before going farther.

"Let us take from the fireplace some very light coals, well calcined, and reduce them to coarse powder. Next let us mix this black powder with a little highly colored vinegar and strain the mixture through a piece of very fine linen or, better still, through filter-paper placed in a funnel. The linen, and still more the paper, will retain the charcoal to the smallest particle, the vinegar alone passing through. But what a singular change will have taken place! The vinegar, at first of a dark reddish hue, has become limpid, showing hardly a trace of red; as far as color is concerned it looks almost like water. But it has lost none of its other properties; its pungent odor and strong taste are the same as at the beginning. Only the color has disappeared. This experiment teaches us something of great interest: charcoal has the property of bleaching liquids by taking to itself the coloring matter contained in them.

"This property is carried to its highest development in a charcoal made from the bones of animals and called for that reason animal charcoal or boneblack. Filtered through this substance in powdered form, vinegar and red wine become as colorless as water, without losing any of their other properties. A few words will tell you how this curious charcoal is made that takes the color out of liquids so easily. Throw a bone on the fire: soon you will see it flame up and turn quite black. If you waited too long, what is now charcoal would be burnt up entirely and the bone would in the end become quite white. But withdrawn before being wholly burnt up, it is as black as common charcoal. Reduce this charred bone to powder, and you will have real bone-black.

"Well, it is by means of half-burned bones, boneblack in fact, that sugar is refined. Bones of all kinds of animals, refuse from the slaughter-house, kitchen remnants, carcasses found in sewers, all are carefully gathered up and converted in kilns to boneblack for the bleaching of sugar until it assumes the whiteness of snow."

"That, then," said Emile, "is what started my friend's jest. Sugar is not made of dead people's bones, but bones turned into charcoal are used to whiten it."

"Yes," his uncle agreed, "that undoubtedly accounts for your friend's odd notion."

"If it were not for their being burned in a hot fire in the first place," Emile continued, "I should be disgusted to think of bones picked up anywhere being used in sugar-refineries. Fire purifies them; otherwise I should stop eating sugar."

"Banish all repugnance on that score, my child. These bones are so thoroughly calcined that not the slightest trace of their former impurity remains. Let me tell you how they are used. The brown sugar, be it from cane or beet-root, is dissolved in hot water and the syrup thus obtained mixed with the proper quantity of bone-black, which draws to itself the impurities that give raw sugar its yellowish color and unpleasant taste. This mixture is strained through thick woolen cloths which act as filters. The charcoal remains above with all the impurities it has contracted, while the syrup passes through as limpid as the water that gushes from a rock. The sugary liquid is then boiled down and finally poured into cone-shaped molds, where it hardens into sugarloaves of irreproachable whiteness and flavor."

Chapter Thirty-Four - Tea

"HAVE you ever examined carefully the grounds in the bottom of a pot of tea? A pinch of tiny round bluish-black grains is put into hot water; after steeping these round grains are found to have turned into easily recognizable little leaves."

"Yes, indeed," Marie made haste to reply, "I have seen the grains of tea swell in hot water, unfold, and finally spread out into little leaves. Tea must come, then, from the foliage of some sort of plant."

"You are right: tea is the leaf of a shrub of which this picture will give you an idea. It is an evergreen shrub, two meters or more in height, its foliage tuftlike and shiny, its flowers white, and its seeds in the form of small capsules in clusters of three. Its cultivation is confined to China and Japan. [1]

"In China the tea plantations

Branch with Flowers of Tea

a, leaf, showing the nervation; *b*, seed-vessel in process of opening; *c*, a seed.

92

occupy sunny hillsides in the vicinity of watercourses. The leaves are gathered, not by the handful, but one by one and with the utmost care. Minute as such work appears, it is done rapidly by trained hands, one person being able to pick from five to six kilograms a day. The first picking occurs toward the end of winter, when the buds open and let the nascent leaves expand. This harvest, considered the best of all, is called imperial tea, being reserved for the princes and rich families of China. The second picking takes place in the spring. At this time some of the leaves have finished growing, while others have not yet reached their full size; nevertheless they are all gathered indiscriminately and then picked over and assorted according to their age, dimensions, and quality. The third and last picking is made toward the middle of summer, when the leaves are of tuftlike appearance and have attained their full growth. This is the coarsest and least-esteemed kind of tea. When the harvest is over, its completion is celebrated by public festivals and rejoicings."

"The harvesting of this leaf must then be a very important event to the Chinese," observed Claire.

"Yes, because tea is the customary drink of the Chinese, being to them what wine is to us; and tea also furnishes them one of their most important articles of commerce. Isn't that reason enough for public rejoicing, especially in a country where everything pertaining to agriculture is held in high honor?

"Before taking the form of tea as we know it, the leaves have to undergo a certain preparation. This work is done in public establishments where there are little furnaces about a meter high, on each of which is placed an iron plate. When the plate is hot enough the operator spreads the newly gathered leaves on it in a thin layer. While they are shriveling and crackling in contact with the burning iron, they are stirred briskly with the naked hand until the heat can no longer be endured. Then the operator removes the leaves with a sort of fan-shaped shovel and throws them on to a table covered with mats. Around this table sit other workers who take the hot leaves in small quantities and roll them between their hands, always in the same direction. Still others fan them continually after they are rolled, so as to cool them as quickly as possible and thus preserve the curled shape they have just received. This manipulation is repeated two or three times in order to drive out all moisture from the leaves and give them a permanent curl. Each time the plate is heated less and the drying is conducted more carefully and slowly.

"The use of tea spread to Europe toward the middle of the seventeenth century. It is said that about this time some Dutch adventurers, knowing that the Chinese made their customary drink from the leaves of a shrub grown in their country, took it into their heads to carry them a European plant, sage, to which great virtues were attributed in those days. The Chinese accepted this new article of commerce and in exchange gave them some tea, which the Dutch took back to Europe. But the use of the European herb was of short duration in China, whereas tea was so highly appreciated in Europe that it soon came into general use.

"There is a tradition in China much like the one current in Arabia concerning coffee. According to this tradition, a certain pious and noble personage, Dharma by name, went from India to China thirteen or fourteen centuries ago to spread the knowledge of the true God in that country. In order to stimulate the people by his own example he led a very austere life, imposing the severest mortifications on himself and consecrating his days and nights to prayer. Worn out by fatigue after a few years, and finally overcome by drowsiness, it sometimes happened that, in spite of himself, he would fall asleep in the very midst of his meditations. In order to keep himself awake and continue his pious exercises without interruption, he had recourse to the frightful expedient of cutting off his eyelids, which he threw on the ground. Heaven was moved to pity by this heroic sacrifice: the holy man's eyelids took root in the soil as if they had been seeds, and there sprang from them during the night a graceful shrub covered with leaves. That was the first tea-plant. The next morning, passing by the same place, the mutilated holy man glanced down at the spot where he had thrown his eyelids. He could not find them, but in their place he saw the divine shrub to which they had given birth. A secret inspiration prompted him to eat of the leaves of this miraculous shrub, and he obeyed the impulse. To his great satisfaction he soon found that this nourishment strengthened him, drove sleep away, and kept his mind active. He advised his disciples to eat of the shrub also, and the fame of tea spread far and wide, its general use in China dating from that time.

"I need not tell you that this tradition is really only a fable emphasizing the dominant properties of tea, just as the Arabian legend concerning the capering of goats and the wakefulness of the dervish is based on those of coffee. A shrub sprung from the eyelids which a holy man had cut off in order not to succumb to sleep ought, above all, to prevent sleep. Tea shares this singular property with coffee. An infusion of its leaves acts on the nerves when it is taken strong and in considerable doses. Taken in moderation, it is an agreeable drink, stimulating the stomach and aiding the process of digestion.

"The various kinds of tea known to commerce are classed, according to the size of their grains, as pearl teas and gunpowder teas, the former having larger grains than the latter. They are divided again according to color into green teas and black teas. Green teas have a bitter and pungent taste and a strong odor, excite the nerves and prevent sleep. Black teas do not have this property in so pronounced a degree, being less stimulating, weaker, and not so strongly scented. The preparation of tea calls for the same care as that of coffee: it should not be boiled, as that would dissipate the odor and take away the crowning excellence of the beverage.

"With us tea is hardly more than a medicine used to alleviate certain stomach troubles; but in many countries besides China it is a daily drink, appearing on the table several times in the twenty-four hours. In England, the European country most addicted to this drink, the annual consumption amounts to twenty-five million kilograms."

Since 1876 tea has been grown in increasing quantities in Ceylon, Natal, Brazil, and the West Indies; and some of our own southern States have also tried, with varying success, to raise the plant. — *Translator.*

Chapter Thirty-Five - Chocolate

"IN the hottest countries of the two Americas, notably in Mexico, the Antilles, and Guiana, there is cultivated a tree of about the size of our cherry tree, called the cacao or chocolate tree."

"What a queer name that is — cacao!" Claire exclaimed; "not a bit like any of our fruit-trees."

"This queer name has come down to us from the primitive inhabitants of Mexico, a people who tattooed their red skin with horrible designs and wore their hair standing up in a menacing tuft adorned with hawks' feathers. Their language was composed of harsh guttural sounds which to our delicate ears would seem more like the croaking of frogs than the speech of human beings. You have a sample in the name of the tree I have just mentioned. The Mexicans, when the Spanish visited them for the first time under the lead of Fernando Cortez, soon after the discovery of America by Columbus, were devoting careful attention to the cultivation of the

Fruiting Branch of Cacao

cacao tree, from which they obtained their chief article of food, chocolate."

"The same chocolate that is used for making those delicious tablets we all like so much?" asked Jules.

"The same, at least as far as the essential ingredients are concerned. We owe the invention of chocolate to the ancient savages of Mexico, ferocious Indians who honored their idols by offering them human victims whose throats they cut with the sharp edge of a flint. The tree that furnishes the chief constituent of our chocolate confectionery is the cacao, the name of which sounds so harsh to your ears.

"This tree grows, as I said, to about the size of our cherry-tree. Its leaves are large, smooth, and bright green. Small pink flowers grouped in little clusters along the branches are succeeded by fruit having the shape and size of our cucumbers, with ten raised longitudinal ribs as in melons. These cacao-pods, as they are called, turn to a dark red when ripe. Their contents are composed of soft white flesh, pleasantly acid, in which are embedded from thirty to forty seeds as large as olives and covered with a tough skin. Freed from all these wrappings, the seeds take the name of cacao-nibs and constitute the essential ingredient of chocolate.

"Much as in the case of coffee, cacao (also called cocoa) is first roasted, a process that turns the white kernels to a dark brown. That is the origin of the brown color of chocolate. After roasting, the hard skin that covers the kernels is broken up and thrown away; then the kernels themselves, first thoroughly cleaned, are crushed on a very hard polished stone with the aid of another stone or an iron roller. These kernels are rich in fat somewhat resembling our ordinary butter, and hence called cacao-butter."

"There is butter in those seeds, real butter such as we get from milk!" asked Claire.

"Yes, my dear, real butter, or something very similar. Of what do the cow and the sheep make the butter that we get from their milk! Evidently of the grass that they eat. What wonder is it, then, that vegetation should be able to produce butter if it can supply animals with the materials for butter? I hope to come back to this subject some day, and you will see that in reality plants prepare the food that animals give us.

"But let us return to cacao-butter. To keep this fatty substance fluid and thus facilitate the working of the paste, it is customary to place live coals under the stone on which the seeds are being crushed. With a little heat the vegetable butter melts and forms, with the solid matter of the seeds, a soft brown paste that can be easily kneaded. With this, paste is mixed, as carefully as possible, an equal weight of sugar, then some flavoring extract, usually vanilla, to give aroma to the product; and the work is done. There is nothing further needed except to mold the still soft chocolate into cakes.

"Such is the composition of chocolate of superior quality. But for the cheaper grades demanded by the trade it is customary to mix in certain ingredients of less cost than cocoa, as for example the starchy constituent of potatoes, corn, beans, and peas. It is even said — but my faith in the honor of the manufacturers makes me hesitate to believe it — that there are so-called chocolates in which not a particle of cocoa is present. Sugar, potato flour, fat, and powdered brick are said to be the ingredients."

"And that horrid trash is sold?" asked Marie incredulously.

"Yes, it is sold; its low price attracts purchasers."

"If they offered it to me for nothing I wouldn't take it," Claire asserted. "What a queer thing to eat — a cake of brick!"

"It is never true economy to buy very cheap things. The manufacturer and the merchant must make their profit. And yet the buyer is always trying to beat down the price. So what does the manufacturer do? He substitutes something worthless for a part or all of what has real value, and then sells his goods at whatever price you please. He gives you something for your money, it is true; but oftener than not you are outrageously cheated. You have, let us say, only a penny to spend on a cake of chocolate; you will get the chocolate, but it will contain very little cocoa, or none at all, a great deal of potato flour, and perhaps some powdered brick. You think you have driven a sharp bargain; in reality you have been sadly duped. For your penny you could have bought several potatoes, which would have been a far better investment, and the powdered brick besides, if you really care for that sort of thing. Always be suspicious of marked-down goods, my children; the low price is low only in appearance and much exceeds the real value of the goods."

Chapter Thirty-Six - Spices

"WHEN we speak of spices we mean those vegetable substances of aromatic odor and hot and pungent taste that are used to heighten the flavor of food and aid digestion. The principal ones are pepper, clove, cinnamon, nutmeg, and vanilla.

"Pepper is the fruit of a shrub called the pepperplant. You have often seen those little round black grains, with such a pungent taste, that are used for seasoning certain kinds of food — sausages for example. Those are the berries of the pepper-plant just as the bush produces them."

"And do those grains, when they are powdered, give pepper?" asked Plant with Berries Jules; "such as we see on the table every day beside the salt-cellar?"

"Exactly," replied Uncle Paul. "The culture of the pepper-plant is successful only in the hottest parts of the world, chiefly in two of the Sunda Islands, Sumatra and Java. It is a shrub with a slender and flexible stalk having the form of a runner and winding around neighboring tree-trunks. Its leaves are oval, leathery, and shiny; its blossoms small, grouped in long and slender hanging clusters; and its berries, which are no larger than our currants, are first green, and finally red when ripe. Pepper is gathered when the

Branch of Pepper Plant with Berries

bunches begin to turn red; The harvested berries are put to dry on mats in the sun, whereupon they soon turn black and wrinkled, taking thereafter the name of black pepper.

"As their sharp flavor is chiefly confined to the outside integument of the berry, this is sometimes stripped off in order to obtain a less pungent variety of pepper. For this purpose the freshly gathered berries are soaked in water, which makes them swell and crack their outer skin. After that they are exposed to the sun and, when dry, all that needs to be done is to rub them between the hands and then fan them to blow away the exterior covering. This process gives white pepper, which is much milder than the black.

"If you examine somewhat attentively a single grain of the spice called cloves, after letting it soak some time in water until it becomes swollen and expanded, you will easily perceive it to be a flower. Cloves are, in fact, the blossoms of a tree called the clove-tree, gathered and dried in the sun before they are full-blown. The upper part of these blossoms, being rounded like a button, bears some resemblance to the head of a nail; the lower part, which is long and slender, is not unlike the pointed portion. From this rough resemblance comes the name of *clove* (which is connected with the Latin *clavus* and the French *clou,* meaning a nail).

"The Moluccas, or Spice Islands, are the home of the clove. It is a fine tree, about fifteen meters high, with slender branches, oval and shiny leaves, and very strong-scented flowers grouped in clusters.

"Cinnamon is the bark of a tree, the cinnamon tree, originally from the island of Ceylon, but now

cultivated in our tropical colonies.

Flowering Branch of Clove Tree and Unopened Bud

With the point of a pruning-knife the bark of the branches is detached in strips, which are laid together according to size, a narrower on a wider strip, and are then exposed to the sun, whereupon they curl up like quills, one within another, in the process of drying."

"From the look of cinnamon as it is sold by the grocer," said Marie, "you can easily see that it is a bark; but I didn't know what country it came from or what tree produced it."

"From the Molucca Islands, noted as the chief source of the world's supply of spices, we get, in addition to cloves, nutmeg, which is now successfully raised in our colonies. The nutmeg plant is a graceful tree which grows nearly ten meters high. In its rounded head and thick foliage it resembles the orange tree. Its leaves are large, oval, glossy green on the upper side, and whitish underneath; its blossoms, small, bell-shaped, and pendent like those of the lily of the valley, are very sweet-smelling. The fruit, as large as a medium-sized peach, is composed of

Flowering Branch of Cinnamon Tree

98

three parts. First comes a fleshy, edible exterior which at maturity breaks in two; next to this is a network of slender strands, very bright scarlet in color, which yields the spice known as mace; and finally, in the center, lies the seed, or nutmeg proper, which is used as spice. This latter is an oval-shaped body of the size of a large olive, its flesh scented, oily, and very firm, with reddish veins running through it.

Flowering Branch of Vanilla Plant

a, the fruit.

"Vanilla grows in damp and shady forests in the coast districts of Guiana and Colombia. It is a plant with a slender stalk that takes the form of a runner and interlaces the neighboring branches, stretching even from one tree to another, and resembling a small cord covered with beautiful green leaves. Its flowers are large and graceful in shape, white inside, and greenish yellow outside. The fruit, the part to which we give the name vanilla, is sought for its balsamic, sweet odor and its mild, very agreeable taste. It is composed of a viscous pulp and a multitude of very small seeds. In shape and appearance it is long and cylindrical, black, slightly curved, and of the size of one's finger. Vanilla is used to flavor custard, whipped cream, and other similar dishes of which you children never refuse a second helping."

Chapter Thirty-Seven - Salt

"SALT, so necessary for the seasoning of our food, is also very useful as a preservative. The pork stored away for the winter's use is commonly salted or smoked, or both, to keep it from spoiling. Beef, too, is salted down, especially as an article of food for sailors on long voyages; and vast quantities of fish — cod, herring, haddock, and mackerel — are preserved with salt and sent to all parts of the world, even to the smallest villages remote from the seacoast. From these various uses to which it is put you will readily perceive that common salt is one of the most valuable of substances.

"But if we judged of the usefulness of a substance from the price it commands in the market, we should fall into the gravest of errors. For example, the diamond takes highest rank in respect to price, a price that is nothing short of exorbitant, but for real use to man, except as an instrument for cutting glass — and as such it is commonly employed by glaziers — it stands very low in the scale. On the other hand, iron, coal, and salt are among the cheapest of substances, the price per pound being considered, while at the

99

same time they are infinitely more useful than the precious stones, which most often serve only to gratify a foolish vanity. Providence takes no heed of this false valuation, but has assigned the highest importance to iron, coal, and salt by scattering them in profusion all over the earth, and a very inferior importance to the diamond by relegating it to some few remote districts in little-known lands, and that too in very small quantities.

"Accordingly, salt, like all supplies required by mankind in general, is very abundant. The sea, covering as it does three quarters of the earth's surface, the sea, of such tremendous depth and volume, holds in its measureless immensity an enormous mass of salt, since each cubic meter contains nearly thirty kilograms. If all the oceans should dry up and leave behind their saline contents, there would be enough salt to cover the whole earth with a uniform layer ten meters thick."

"What is the use of all that salt!" asked Marie.

"Its use is to preserve the ocean waters from corruption despite all the foul matter therein deposited by the countless denizens of the deep and in spite of the impurities of every kind unceasingly poured in as into a common sewer by the rivers, those great scavengers of the continents."

"They say sea-water is undrinkable, "remarked Claire.

"I can well believe it," assented her uncle. "In the first place, it is very salt, and then it has an acrid, bitter taste that is unbearable. A single mouthful of this liquid, clear and limpid though it is, would produce nausea. Hence it cannot be used in preparing our food, since it would impart its own repulsive flavor; nor can it be used for washing clothes, because soap will not dissolve in it and, more than that, the clothes in drying would retain an infiltration of salt just as does the codfish you buy at the grocer's.

"I have already described [1] to you how salt is gathered from salt-marshes with the help of the sun's heat to dry up the water and leave the crystallized salt ready to be scraped together and carried away. Indeed, the sea is an inexhaustible reservoir of salt: we could never get to the end of it, however lavishly we salted our food. To supplement this abundance, the soil itself, the earth, contains in its depths thick beds of salt which are worked with pick and drill just as stone for building is worked in the quarry. This salt that is dug out of the earth is called rocksalt. It differs from sea-salt only in its color, which is due to various foreign substances, being most often yellow or reddish, sometimes violet, blue, or green. When intended for table use or cooking, it is purified with water, and then is undistinguishable from sea-salt.

"There are salt-mines in the departments of Meurthe and Haute-Saone, but the greatest salt mine is that in the neighborhood of Cracow in Poland. Excavations have there been made to the depth of more than four hundred meters. The length of the mine exceeds two hundred leagues, and its greatest width is forty leagues.

"In that bed of salt are hewn out great galleries with loftier vaults, in some instances, than that of a church, and extending farther than the eye can reach, crossing one another in every direction, and forming an immense city with

100

streets and public squares. Nothing is lacking to the completeness of this subterranean town: divine service is held in vast chapels cut out of the solid salt, and dwellings for the workmen, as well as stables for the horses employed in the mine, are likewise hewn out of the same material. There is a large population, and hundreds of workmen are born and die there, some of them never leaving their underground birthplace and never seeing the light of the sun. Numerous lights, constantly maintained, illumine the city of salt, and their beams, reflected from crystalline surfaces on every hand, give to the walls of the galleries in some places the limpid and brilliant appearance of glass, and in others cause them to shine with the beautiful tints of the rainbow. What magic illumination in those crystal churches when a thousand candles are reflected by the vaulted roof in gleams of light of all colors!"

"Yes, it must be a magnificent sight," Jules assented; "but, all the same, I should want to come up now and then into the light of day."

"Undoubtedly; for with all its splendors that subterranean abode is far inferior to ours. We have the open air, that pure air with which we delight to fill our lungs; and we have the sunlight, a vivifying light that no artificial illumination can equal."

"Nevertheless I should like to see that mine," said Emile. "What a tremendous grain of salt, to hold whole towns!"

[1] See "The Story-Book of Science."

Chapter Thirty-Eight - Olive Oil

"OIL is obtained from seeds of various sorts and from certain kinds of fruit, but the most highly esteemed oil for the table, the very queen of oils, is that which we get from the olive, the fruit of the olive-tree. This precious tree, which the ancients made the symbol of peace, fears the rude winters of the North and thrives with us only in Provence and Languedoc, especially in the departments bordering on or near the Mediterranean. In size it is not a tall tree, usually attaining about twice the height of a man. Its head is rounded, not very dense in growth, and furnishing but poor shade; its leaves are narrow, leathery, of an ashen-green color, and do

Olive
1, Branch with fruit. 2, Branch with flowers. a, A flower.

not fall in winter. In summer its sparse branches are the favorite resort of grasshoppers, which, reposing on the bark of the tree in luxurious exposure to the intense heat of the sun, indulge in unrestrained exhibition of their musical accomplishments.

"The olive is green at first. The flesh covering its hard stone, which is pointed at both ends, has the most disagreeable taste you can imagine. An unripe grape is sour, an immature pear harsh, a green apple tart; but an olive not yet fully ripened far surpasses them all in its repulsive flavor. At the very first bite its unbearable acridity bums the mouth so that you might think you were chewing a red-hot coal. Certainly he had need of a rare inspiration who first reposed confidence in this unpleasant fruit and succeeded in extracting its oil, which is mildness itself."

"I once took a notion," said Marie, "to taste of an olive as it grew on the tree, and I can tell you I soon had enough of it. Goodness, what a horrid fruit! How can such sweet oil come from such bitter flesh?"

"Later when the cold weather of approaching winter comes, in November and December, olives change from green to reddish, and finally turn black. Then the skin wrinkles and the flesh ripens, losing its tartness and becoming rich in oil. That is the time for harvesting the fruit. Women with the help of short ladders gather it by hand and fill their upturned aprons, blowing now and then on their fingers, benumbed by the piercing cold of the December mornings. The harvest is piled up at the foot of the olive tree on a cloth that has been spread there, and the picking is resumed amid interminable chattering and bursts of laughter from among the branches.

"The olives are taken to the mill, where, after being crushed under vertical millstones, they are cold-pressed. By this first pressure is obtained fine or pure oil, the most esteemed of all. Subjected to the action of hot water and pressed a second time, olives furnish a second grade of oil. Finally the residue, mixed with the imperfect olives, such as windfalls and those that are worm-eaten, yields what is known as pyrene oil, which is too ill-flavored to be used in cooking, but is useful for lighting and for soapmaking. The very last residue is made into oilcakes, an excellent fuel."

"But it isn't their oil alone that makes olives valuable," said Marie; "they are good to eat after some sort of treatment that I should like to know about."

"Olives that are black, very ripe, and wrinkled, can at a pinch be eaten just as they come from the tree, in spite of a slight harshness of flavor that still clings to them. To remove this they are slightly salted, sprinkled with a few drops of oil, and kept in a pot, where they are stirred from time to time. In a few days they are ready to eat. Sometimes they are merely soaked in salt water.

"But however they may be prepared, black olives are never equal to green ones. The most ill-flavored olives as they hang on the tree are the best when once freed of their extremely disagreeable taste. Energetic treatment is necessary to give them the desired mildness of flavor. Recourse is had to potash, that harsh substance I told you about in speaking of ashes and the use of lye in washing. A quantity of very clean ashes is taken from the fireplace and put into water, to which is added a little lime, the effect of which is to increase the strength of the potash. Finally the clear liquid, charged with the soluble portion of the ashes, is poured over the green olives. After some hours of contact with this corrosive fluid they lose their acridity, and all that remains to be done is to rid them of the lye that impregnates them. This is accomplished by soaking them in pure water and changing the water every day until it is colorless and tasteless. By this repeated soaking nothing that the ashes had contributed is left. Lastly the olives, now of a beautifully green color and an agreeable taste, are salted down in brine, which insures their preservation and corrects any undue sweetness of flavor."

"Then it is potash," said Claire, "That turns the horrid-tasting fruit into the olives we see on the table, and that I am so fond of."

"Yes, it is the potash obtained from ashes, potash alone, that subdues and softens the harsh flavor of the olive. Add this service to those that the same substance renders us when it enters into the composition of soap, glass, and washing fluid.

"In the case of the olive that furnishes the oil; but other forms of vegetation valued for their oily constituent have this in their seeds, their kernels. Leading examples are the walnut, sesame, poppy, colza, a kind of turnip called rape, and flax. Break a dry walnut, take a quarter of the meat, and hold it close to the flame of a lamp. You will see it

Flowering Branch of Walnut Tree, with Fruit

catch fire and burn with a beautiful white flame which feeds on an oily juice that oozes out as the heat increases. Thus we find there is oil in walnuts. To extract it, we crack the nuts and take out the meats, which we subject to strong pressure. Freshly made nut-oil is pleasant to the taste and well adapted to culinary purposes; it is much in demand wherever nuts abound. Unfortunately, it soon turns rancid and it contracts with age a strong and exceedingly repulsive taste.

"Sesame is an herbaceous annual cultivated chiefly in America and Egypt. Its seeds furnish a sweet oil having nearly the same qualities as olive-oil. Sesame-oil is not sold as such with us, but I sus-suspect dealers occasionally mix it with olive-oil, which is much more expensive.

"Poppy-heads are full of very fine seeds which furnish a fairly good oil known as poppy-oil."

Sesame

"Poppy-heads will put a person to sleep," observed Claire. "I remember once somebody made me drink some poppy-tea to put me to sleep when I was ill. Poppy-oil ought to make one sleep too."

"It is perfectly true that poppy-heads make a tea that induces deep sleep. They owe this property to a substance called opium, which is so powerful that if you took a dose of it no bigger than a pea it would act as a deadly poison and put you to sleep forever. But this formidable substance is found only in the shell of the fruit, in the envelop of the poppy-head, and not in the seeds. Hence the oil extracted from these seeds can be used in cooking without any danger.

Poppy

a, upper part of stem with flower; *b*, lower part of plant; *c*, the fruit.

"Colza and rape are two varieties of turnip cultivated principally in the North. The pod-like fruit contains two rows of fine seeds under two long strips or valves that open from bottom to top at maturity. These seeds give colza-oil and rapeseed-oil, which are used for lighting and also in some of the industrial arts, but are unsuitable for cooking on account of their offensive taste.

"Linseed-oil, finally, which is used chiefly in painting, I have already told you about in one of our former talks."

Chapter Thirty-Nine - The Double Boiler

"To avoid the danger of scorching or 'burning on' in cooking food over a fire, the double boiler has been invented. This insures an equable temperature in the process of cooking, with no risk whatever that the temperature will rise above the boiling-point of water, no matter how hot the fire may be under the boiler. Into a kettle partly filled with water is fitted a smaller one with a rim near the top to support it. The inner kettle contains the substance to be cooked, and is provided with a lid. There you have the double boiler. The fire acts directly on the contents of the lower or outer boiler, heating the water, which in turn transmits its heat to the upper or inner boiler. In this manner no excess of heat can by any possibility injure the cooking food, and there is never any crust of burnt matter found clinging to the bottom of the vessel in which the food is contained, simply because no part of that food ever passes the comparatively low temperature of boiling water."

"But if the fire is stirred and more fuel put on," objected Marie, "The water in the under boiler would get hotter and hotter, and the contents of the other surely would burn!"

"No, my child, there you are quite mistaken. The water in a kettle or boiler can never pass a certain limit of heat, let the fire underneath be as hot as you please. Suppose we put a saucepan of water on the fire. The liquid heats gradually and finally begins to boil. Well, when it has once started boiling it is as hot as it ever will be. In vain would you stir the fire and pile on fuel and blow the flame; your energy would be wasted. The water would boil faster, it is true, but it would not become any hotter."

"Do you mean to say," asked Emile, "that a great bed of red-hot coals would not heat water so that it would be any hotter than if there were only a handful of burning sticks under it?"

"It would make it boil faster," was the reply, "and it would generate more steam; but I repeat that the water would not rise a single degree in temperature when once it had begun to boil. Nothing in the world can increase the heat of boiling water so long as the steam is free to escape."

"It is very strange," commented Jules, "That a great fire cannot make more heat than a small one."

"Ah, but you must not fall from one error into another," returned his uncle. "A great fire does, plainly enough, make more heat than a small one; but this excess of heat is not retained by the water, and therefore the latter does not become any the hotter."

"I should like to know," said Claire, "how they make sure that boiling water cannot get any hotter than when it first begins to boil, no matter how hot the fire underneath. They don't put their hands in to find out, I suppose."

"Certainly not. The test is made with a thermometer, the little instrument I have already described [1] to you. If a thermometer is plunged into boiling water the mercury or spirits contained in the bulb and the lower part of the glass tube will be seen to rise until the division marked one hundred (centigrade) is reached — never more, never less, however hot the fire, so long as the water is free from any intermixture of other matter. If the fire burns furiously the water will boil all the faster and will send off great volumes of steam; if the fire is low the water will boil slowly and send off but little steam; but in each case, so long as it boils at all, it will be of exactly the same temperature — one hundred degrees centigrade. In this way it is established beyond question that freely boiling water can never rise above a certain temperature, no matter how hot the fire underneath.

"The usefulness of the double boiler is thus made plain to you. The inner kettle, immersed in boiling water, can never be subjected to a heat greater than that of boiling water. Now, many substances, especially those that serve us as food, suffer no injury from being raised to the boiling-point of water, while they scorch and acquire a bad taste if heated to a higher temperature. Such, for example, is the casein of milk. Therefore in cooking preparations containing milk, it is well to use the double boiler. And in melting butter in order to free it from casein and make it keep better — a subject we have already discussed [2] — the double boiler is better than a single kettle. The casein is deposited at the bottom without risk of burning, and the pure butter is poured off at leisure and stored in pots or jars, which are sealed up and kept from year to year without deterioration of the contents, so far as cooking purposes are concerned."

[1] See "The Story-Book of Science."   [2] See "Our Humble Helpers."

Chapter Forty - Little Pests

THE chief constituent of cheese, as we have already seen, [1] is casein, which is separated from the rest of the milk by the action of rennet. But casein alone would make very poor cheese, and therefore it is customary to add more or less cream to it and thus furnish a cheese of greater richness and savor. The amount of cream added determines in general the richness and palatability of the cheese, and thus innumerable grades and varieties of this article of food are offered for our selection; yet they all owe their origin to the one substance, milk.

"Kept too long, as I have told you before, all cheeses, some earlier, others later, become moldy, first on the outside, then on the inside, the mold being at the outset of a yellowish white, afterward blue or greenish, and finally brick-red. At the same time the substance of the cheese decays and acquires a repulsive odor and a flavor so acrid as to make the lips smart. Henceforth the cheese is nothing but a putrid mass which must be thrown away. Deterioration is more or less rapid according to the softness or hardness of the cheese and its permeability by the air. To make cheese keep well, therefore, it must be dried thoroughly and reduced to compactness by strong pressure. Certain varieties of Dutch cheese, remarkable for their durability, are so hard and compact that sometimes, before they can be eaten, they have to be broken up with a hammer and softened by wrapping in linen soaked in white wine. But, hard though they are, these cheeses are valued for seasoning, for which purpose they are first reduced to powder on a grater; and they are also serviceable in the provisioning of ships for long voyages.

"Mold and decay are not the only enemies of cheese; there are also certain little creatures, mites and worms, that invade its substance and establish themselves there, defiling the cheese by their presence and gnawing it away, little by little. The cheese mite, or *acarus domesticus*, as it is called by the learned, is a tiny creature hardly visible to the naked eye, with a body all bristling with stiff hairs and supported by eight short legs. Burrowing with its pointed head into the soft cheese, it lives there in colonies of a membership past counting, protected by the rind and taking shelter in the crevices. Assembled in mass, these animalcules look like so much dust, though on closer inspection it is seen to be animate dust, moving and swarming, and resolvable into a prodigious number of extremely small lice. If these mites are allowed to multiply at their own sweet will, the cheese gradually crumbles to dust. To ward off their inroads cheeses should be occasionally scrubbed with a stiff brush and the shelves holding them washed with boiling water. Cheese already attacked should first be well brushed and then rubbed with oil, which kills the mites. A more energetic procedure consists in subjecting the cheese to the fumes of burning sulphur in a closed box or chest. The sulphurous gas kills the animalcules without in the least impairing the quality of the cheese."

"And what about the worms you spoke of?" asked Claire.

"They are even more to be feared than the mites. What could be more disgusting than a piece of cheese promenading, so to speak, across one's plate, borne on the backs of these horrid creatures!"

"Sometimes they are so numerous," remarked Marie, "that the substance of the cheese seems changed into vermin. It must be the decay that turns the cheese into mites and worms."

"No, indeed, my child; never in all the world does decay engender vermin. Cheese-mites and cheese-worms come from eggs laid by other mites and by the flies into which the worms are finally changed just as caterpillars are changed into butterflies."

"Then the vermin that we see swarming in all sorts of decay does not really come from that decay?"

"Surely not. The decay feeds the vermin, but never brings it into being. It comes from eggs laid by various insects, especially by flies that are attracted from a distance by the smell of decay. Thus cheese-worms finally turn into flies of various kinds whose lifetime is spent in the open air, often amid the flowers. When the time comes for laying their eggs these flies, guided as they are by the sense of smell, know very well how to find our supplies of cheese. There they deposit their eggs, each of which becomes a little worm which later turns into a fly."

"How about the worms we find in fruit?" asked Jules. "Do they too come from eggs?"

"Yes, all worms, wherever they may be found, owe their origin to eggs laid by insects; and never, bear in mind, are they produced directly by decay. Let me give you a few examples.

"Who is not familiar with the cherry-worm? The cherry itself may be of fine appearance, plump, dark purple, bursting with juice. Just as you are about to put it to your lips you feel a certain softness near the stem, and your suspicions are aroused. You open the cherry. Pah! A disgusting worm swims in the decaying pulp. That is enough. Those fine cherries tempt you no more. Well, that worm, if left undisturbed, will turn into a beautiful black fly, the cherry *ortalis,* with diaphanous wings crossed by four dark bands. The insect lays its eggs on cherries still green, one on each. No sooner is it hatched than the worm bores a hole through the pulp and installs itself next to the stone.

This orifice is very small and, more than that, it soon closes up, so that the fruit inhabited by the worm looks sound. The worm's presence does not interfere with the growth and ripening of the cherry, a fortunate circumstance for the worm, which is thus allowed to gorge itself with the juicy sweet pulp. When the cherry is ripe the worm is also fully developed, after which it leaves the fruit and drops to the ground, where it digs itself in and waits for the next May, when it will turn into a fly, lay its eggs on the young cherries, and die."

"Now I understand," said Marie, "how the worm gets into the cherry. I had always supposed it came in some way from the decaying pulp of the fruit."

"I will next show you," continued Uncle Paul, "a picture of the insect that in its worm state eats nuts. It is called the nut-weevil. With its long, pointed beak it pierces a hole in the tender shell of the young fruit, and at the bottom of this hole, in contact with the nut-meat, it deposits an egg which in a few days hatches a tiny worm. As this worm eats but very little the nut continues to grow and the nut-meat ripens; but the gnawing goes on. Some time in August the stock of food comes to an end and the worm-eaten nut falls to the ground. The weevil itself, its jaws now robust, bores a round

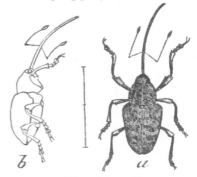

Nut-Weevil

a, back view; *b*, side view. (Vertical line shows natural size, including proboscis.)

hole in the empty shell and abandons its early home to burrow into the ground, where its transformation takes place and the worm becomes a perfect insect."

"I often find," said Emile, "under the hazelnut trees in the garden, nuts that look all right at first, only each one has a little hole in it and no meat inside."

"The meat has been devoured by the nutweevil, and the round hole is the door by which the creature made its exit."

"Sometimes," said Claire, "when I crack nuts with my teeth, I bite into something bitter and soft."

"That," returned her uncle, "is the worm of the nut-weevil, crushed by your teeth."

"Pah! The nasty thing!" she exclaimed.

"And what about the worms we often find in apples and pears?" asked Marie.

"Those are what are commonly called appleworms, or in learned language, *Pyralidae*. The moth has wings in two pairs, the upper being of an ash-g ray marbled crosswise with brown and adorned at the wing-tips with a large red spot surrounded by a golden-red band; the lower of a uniform brown. As soon as the fruit begins to form the insect lays an egg in the blossom-end of the apple or pear, either fruit being alike acceptable. The tiny worm that hatches out, no bigger than a hair, bores into the fruit and establishes itself near the seeds. The little orifice by which it entered soon heals over so that the apple or pear appears sound for some time.

"Meanwhile the worm goes on growing in the lap of plenty. It makes a hole communicating with the outside, to admit fresh air and insure the ventilation essential to the sanitary condition of its abode with all its encumbrance of rubbish and excrement. By this tunnel bored through the pulp of the fruit till the outside is reached the worm both receives fresh air and from time to time ejects, in the form of reddish wormhole-dust, the pulp gnawed and digested. Apples and pears thus infested do not cease to grow; on the contrary they mature even earlier than the others, but it is a sickly maturity, and hastens the fall of the fruit. The worm in the fallen apple or

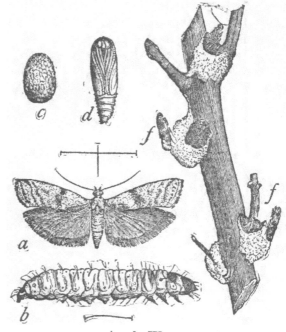

Apple-Worm

a, moth; *b*. larva; *c*, egg; *d*, pupa; *f*, *f*, moths, natural size, resting on egg-sacs of *pulvinaria*. *a* is about twice natural size; *b*. *d*, more enlarged; *c*, highly magnified.

pear leaves its domicile by the exit already prepared and withdraws into a crevice in the tree's bark, or sometimes into the ground, where it fashions for itself a cocoon of silk mingled with bits of wood or dead leaves; and the next year it turns into a moth, at the season that brings forth the young apples and pears in which are to be laid the eggs for a new generation of worms."

[1] See "Our Humble Helpers."

Chapter Forty-One - Flies

"THESE examples," Uncle Paul resumed, "which I could multiply indefinitely without finding a single exception, prove to you that all vermin swarming in decayed matter owes its origin to eggs laid by insects, flies, moths, butterflies, and beetles, of various kinds. Life always springs from life, never from decay."

"All the same," Marie declared, "there are lots of people that say rotting matter breeds worms."

"That is an error as old as the world, and even in our day it is widely disseminated, though much less so than in ancient times. Persons of the highest education used to regard it as beyond question that mud, dust, excrement, and other matter in decomposition would breed animal life, even rather large specimens, such as rats, frogs, eels, snakes, and many others. If the learned men of antiquity endorse in their works such gross errors, what must have been the beliefs of the uneducated!"

"Didn't those learned men know," asked Claire, "What frogs come from tadpoles hatched out of eggs laid by other frogs?"

"They did not know it."

"They had only to look into a pond to find it out."

"They did not know how to look. In those old days men reasoned a great deal too much, sometimes to the point of unreason; but seldom did they take it into their heads to examine things as they really are. Patient observation, mother of all our knowledge, was unknown to them. They said, 'That is it,' without examining the matter, whereas in our day we examine before saying, "That is it.' By this reversal of method science has, in scarcely a century, attained to a degree of power that astonishes us with its marvelous achievements. It is observation that has given us the means of protecting ourselves from the thunderbolt by using the lightning-rod; of covering enormous distances in a short time with the help of steam, which propels the railway locomotive; and of transmitting thought instantly from one end of the world to the other with the electric telegraph. Truth is acquired through observation; man does not invent it, but has to seek it laboriously, and is fortunate if he finds it.

Tadpoles

"For want of close observation the ancients, on seeing a litter of young mice come out of some hole in the wall, attributed the procreation of these animals to the dust of the wall. If they saw a company of frogs leaping about on the muddy banks of a pond, that was enough to make them believe frogs sprang from mud fermenting in the sun."

"And I," declared Emile, "am sure they are hatched out of eggs. From one of these eggs comes first a tadpole, which little by little loses its tail, gets four legs, and finally turns into a frog. That is something like the change of form of a caterpillar into a butterfly."

A, B, with gills; C, more advanced. a, eye; o, ear; m, mouth; n, nasal sacs; d, opercular fold; kb, ki, gills; ks, a single branchial aperture; z, horny jaws; s, suckers; y, rudiment of hind limbs.

"You know what was unknown to a great many wise heads in olden times."

110

"If I know it, it is thanks to Uncle Paul; and those who think worms come from decayed matter and frogs from mud apparently have no Uncle Paul."

"Alas, my child, how many there are that have none! By that I mean there are few who receive the thorough education that enables one to judge things from experience, observation, and sound reason. People trust the merest appearances and transmit their own premature conclusions. It is least troublesome and the quickest way. As you grow older, my dear children, you will learn how many foolish sayings gain currency in the world because people will not take the trouble to reflect and observe — observe with their own eyes.

"If one is but willing to learn, for example, that the worms in decaying matter come from eggs and not from the decay itself, all that is necessary is to have eyes and use them; for the simplest sort of experiment will decide the question, though it was centuries and centuries before any one thought of it. We merely cover with gauze or a fine wire screen any food that is beginning to spoil, such as cheese on the point of going bad, or anything else of the sort. Attracted by the odor, flies soon come circling about these tainted substances, and even lay their eggs on the gauze at the points nearest to the decaying matter which lies just out of their reach. Under these conditions, however far advanced the state of decomposition may be, no worms will make their appearance in the tainted food, because it has been kept where no eggs could be laid in it. But if the protecting gauze or wire screen is removed the flies will lay, here and there on the decaying substance, piles of little white eggs, and very soon there will be thousands of worms swarming amid the decomposing organic matter.

"By means of observations requiring rather more care it is possible to catch in the very act the little insect that lays in the cherry the egg from which comes the worm we all know so well. It has been ascertained that wormy fruit owes the inhabitants that devour it, not to decay as such, but to eggs deposited there by various insects. It has been discovered that lice do not come from flesh, nor fleas from fermenting excrement, and also that frogs are not engendered by pond mud. In short, a thousand errors of this kind have been so completely refuted that there remains not the shadow of a doubt on the manner in which the smallest grub is brought into being. Wherever you find worms, caterpillars, insects, be assured that other insects have laid their eggs there. Always and everywhere life owes its existence to life."

"You open new views to us, Uncle," said Marie, "and they will rid us of ever so many false notions."

"I have merely been trying to show you the salutary part our reasoning faculties are called upon to play. It now remains for me to give you a lesson from established facts. Certain articles of food used by us, such as cheese and meat, especially game, are always in danger of falling a prey to worms. This odious class of vermin owes its existence to flies which, according to their species, go in quest of animal flesh or of cheese wherein to lay their eggs. Two species are already well known to you. for you often see these flies

111

buzzing noisily on the windowpane. The first kind is dark blue, the second grey with reddish eyes. Both of them have much larger bodies than the ordinary fly, and both attack meat. As to cheese-flies, I need only remind you of the worms too common to be unknown to you.

"These flies, these winged pests and audacious parasites — in them you behold the enemy must be kept at a distance and prevented from laying eggs in provisions if we wish to guard our larders from the invasion of vermin. Cut cheese should, accordingly, be kept under a bell-shaped wire screen or, better, under a glass dome, which at the same time insures its protection from flies and keeps it from drying up through prolonged exposure to the air. As to meat and game, which

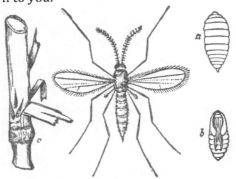

Hessian Fly

a, larva; b, pupa; c, infested stalks of wheat.

need a continual circulation and renewal of air, they should be hung in cages of fine wire netting, and every time the cage is opened care should be taken not to let in any of the blue flies that are usually lurking in the neighborhood. If the enemy were shut up with the provisions even for twenty-four hours, everything would be spoiled, such a multitude of eggs does the blue fly lay, and in so short a time. In a securely closed and carefully watched cage, game, however strong its flavor, will always be free from worms unless they were already in the game when it was placed in the wire cage."

Chapter Forty-Two - The Three States of Matter

"STONE, a piece of wood, a bar of iron, are objects more or less hard which offer resistance to the finger and can be grasped and handled. Cut or chiseled into any desired shape, they will retain that shape. On account of these properties we say of stone, wood, iron, and other substances that resemble them in this regard, that they are solid substances.

"In everyday language this term 'solid' is applied to any object that offers great resistance to rupture, to deformation. For example, we say 'This piece of wood is solid,' 'This iron hook is very solid.' That is not the way the word should be understood in the present connection. I call solid any substance that can be grasped and handled, any substance, in short, that keeps the shape given it. Thus butter, tallow, moist clay, are plastic substances readily molded by the hand into any desired shape. We can grasp and handle them without difficulty, can fashion them as we please. In this sense they are solid substances no less than marble and iron, which are so resistant."

"That is easy to understand, "said Claire. "Anything that can be handled, even if softer than butter, is called solid. So water is not solid, for I can't take up a pinch of it in my fingers as I do with sand. Neither can I shape some of it in the form of a ninepin, for example, and stand it up; or at least I can't unless I put it in a bottle."

"Golden-tongue could not have said it better. No, water is not solid. It slips through the hand that tries to hold it; it flows. Left to itself it has no shape, and it is impossible to give it a definite one except by enclosing it in a vessel. Then it adapts itself to the form of the container, taking its exact shape — round if the vessel is round, cubical if the vessel is a cube. Water and other substances that flow are called liquids."

"Then milk, oil, wine, vinegar, melted butter, are all liquids," said Jules.

"Yes, they are liquids the same as water.

"Now let us turn our attention to the steam that escapes from a boiling pot or, if you like, to the beautiful plume of white vapor that comes in puffs from the smoke-stack of a locomotive as the latter moves along on the iron rails. You remember those magnificent puffs ascending in billows that remind one of the softest kind of swans' down."

"I know what you mean," Emile hastened to reply; "the engine puffs them out with a loud noise like a person blowing with all his might."

"Well, those white puffs are steam from water, just like the white puffs from the little boiling pot. This steam makes the locomotive move, and then, after it has done its work, it escapes with a loud noise into the air. Here we have another substance impossible to grasp; and this impossibility is greater even than in the case of water. Handling it is quite out of the question. Moreover, it expands in all directions, gaining in volume and occupying an increasing amount of space. On issuing from the smoke-stack the puff of steam had a certain volume, not very large. Inside the engine it had still less, and that is precisely what gave it its force; for, like a spring that possesses more energy the more it is pressed down, steam owes its power to the fact of its confinement within a restricted space. Once set free, it gains more and more volume until at last it becomes so dispersed as to be invisible. You must, in fact, have noticed that the white plume soon melts, as it were, in the air and disappears.

"Invisible though it thus becomes, it is clear that this steam exists and that it belongs to a special class of material substances. Is not air itself intangible and invisible! And yet can one doubt its materiality when, as wind, it is set in violent motion and makes the trees rock and sway, or even tears them up by the roots I Thus we perceive there are substances characterized by an extreme thinness, the thinness of the air we breathe. These substances do not retain any fixed form like solids; they have no constant volume like liquids; they expand in all directions and, unless confined, occupy more and more space. They are called aeriform substances on account of their resemblance to air; they are also known as gases and vapors. Air is a gas. To this class belong also the invisible but pungent fumes of burning sulphur, and the green-

113

ish substance of unbearable smell whose properties I described to you in our talks on coloring matter and on ink in particular. The first-named substance is sulphurous oxide, useful in bleaching wool and silk; the other is chlorine. Lastly, the invisible steam from boiling water is also a kind of gas, or rather a vapor, for gas and vapor are much alike."

"And that kind of air full of something that comes from burning charcoal, that dangerous air that gives ironers a headache if they are not careful to keep their heaters under a chimney — that must be a gas too?" This query came from Claire.

"The deadly substance emitted by burning charcoal is in truth a gas, as invisible and as odorless as air itself. It is called carbonic oxide.

"Thus all substances, or, to use another term, all bodies, assume one or other of the three different forms known as the three states of matter; namely, solid, liquid, and gaseous.

"Now the same substance can, without changing its nature in the least, become in turn solid, liquid, and gaseous, according to circumstances. It is mainly heat that effects these transformations. Heated to the requisite temperature, certain solids become liquid; with still more heat the liquid becomes a gas. In losing heat, on the other hand, that is to say in cooling, a gaseous body passes successively from the gaseous to the liquid state, and from that to the solid. The following example will show this more clearly than any mere description.

"Ice is a solid body; many stones are no harder. Let us put it on the fire in a vessel. It will melt; in gaining heat it will become liquid water. If this water in its turn is heated still more, it will begin to boil and will pass off in vapor; that is to say, it will take the gaseous state. Here, then, we see water changing, under the action of heat, from the solid to the liquid state, and from the liquid to the gaseous. Most bodies are subject to similar changes. It is true that sometimes heat of extreme violence is needed; thus iron will not melt unless subjected to the intense heat of the blast-furnace; and to vaporize the smallest particle of it requires the most tremendous sort of fire that science can produce. And so with varying degrees of reluctance all elemental substances obey this common law: heat first melts them, makes them become liquid, then volatilizes them, that is to say reduces them to vapor.

"What does cold do on its part? First take notice that cold has no real existence, that it is not something opposed to heat. All bodies without exception contain heat, some more, some less, and we call them hot or cold according to whether they are warmer or colder than we. Thus heat is everywhere, and cold is only a word that serves to designate the lesser degrees of heat. To cool a body is not to add cold to it, there being no such thing as cold; it is taking heat away. If a body gains heat it becomes warm; if it loses heat it turns cold.

"Well, then, the act of cooling, that is to say the withdrawal of heat, restores vaporous bodies to the liquid state, and liquids to the solid state. Thus the steam from the boiling pot on coming in contact with the cold lid loses its

114

heat and turns to water again; and the vapor in our breath, when it touches a pane of glass, becomes cold and runs down in fine drops. Water in its turn if sufficiently cooled turns to ice, that is to say becomes solid. Other substances act in the same way: a diminution of heat brings them back from the gaseous to the liquid state, then from the liquid to the solid."

Chapter Forty-Three - Distillation

By the action of heat liquids are vaporized, and the vapor in its turn becomes a liquid again on cooling. Suppose, now, there is a mixture of two liquids of which one turns to vapor more easily than the other. On the application of heat, with the exercise of a little care, the more easily vaporized of the two liquids will be the first to evaporate; and if this vapor, instead of being allowed to escape into the air, is held confined in a cool receptacle, it will return to the liquid state. In this manner the two mingled liquids will be separated, the one less easily vaporized remaining in the vessel used for heating them, the other being collected in another by itself. This operation of separating the two is called distillation.

"To fix this process well in mind, let us take an imaginary example. Let us suppose we have a quantity of water into which has been poured a considerable portion of ink. The liquid therefore is black, unfit for drinking and unfit for any other purpose requiring the use of water."

"But who would ever dream of drinking water as dark as shoe-blacking," Claire interposed, "or of using it to wash linen or cook vegetables?"

"Nevertheless let us see if there isn't some way to restore this water to its original purity, to separate it from the dark pigment in the ink and make it as clear and limpid as ever. Yes, there is a way: it is the method adopted in distillation. Water is easily vaporized, whereas the coloring matter in the ink is vaporized with extreme difficulty. If, then, we apply heat the water alone will rise in the form of vapor, while the dark matter will remain behind. Thus heat will bring about a separation that at first seemed impossible. All we have to do now is to confine the water-vapor and cool it until it returns to the liquid form; then the process is complete and we shall have in one vessel perfectly clear water, in another a turbid liquid containing the ink."

"If you had asked me," said Clair, "to separate the two, the water and the ink, after they had once been mixed, I should have said it couldn't be done. And yet how easy it is! We heat the mixture and the separation takes place of itself. I should like to see this curious experiment."

Retort (*a*) and Receiver (*b*)

"Nothing would be easier than to show it to you if we had the necessary apparatus. All that I can do at present is to show you a picture here that will help to make the process clear. We put the darkened water into a glass vessel called a retort, which expands at one end into a large globular flask, and at the other contracts into a long, tapering neck. The flask of the retort, when in action, is placed over a fire or flame."

"Can glass be used for boiling water?" asked Jules.

"Certainly, if it is thin enough to expand uniformly when placed over the fire. The glass in this instance is of a quality that will bear heat if proper care is exercised in conducting the operation. Owing to its transparency it affords a clear view of what takes place inside, a circumstance of great importance when we desire to follow the successive steps of an experiment. The neck of the retort is inserted into another receptacle, likewise of glass and globular in shape, which is plunged into cold water. If heat is applied beneath the flask the water contained in it is vaporized, while the coloring matter is not. This vapor, as fast as it reaches the cooling receptacle, immersed as the latter is in cold water, loses its heat and returns to the liquid state. Thus we obtain perfectly clear water, free from all traces of ink. Spring-water is not clearer or purer; indeed, it is less pure, as you will presently perceive."

"That's all very clever and very interesting," observed Jules, "to be able to get clear water out of a bottle of ink; but what's the good of it? No one would ever think of such a thing as blackening water with ink just to turn it back into clear water by distilling it."

"Very true," was the reply. "I chose that example in order to make the process more striking to you. But if it is not our practice to obtain pure water for daily use by distilling it from a mixture of ink and water, it is no unusual thing to distil ordinary water, and for this reason: however clear and good to drink water in its natural state may be, it is never strictly pure. Whether it comes from a well, a spring, a river, or a lake, it has been in contact with the earth and consequently must contain, in however small a quantity, some of the soluble constituents of the soil. Would not water be salt if it ran over a bed of salt, and would it not be sweet if it ran over a bed of sugar? In like manner water that washes the soil is charged with the numberless soluble substances contained therein. Who has not noticed the earthy deposit left in course of time by even the best waters on the inside of bottles and pitchers and, in a still more marked degree, of water-pipes? What is this deposit except an incrustation gradually formed by the foreign substances dissolved in the water? No water, then, that comes in contact with the soil is pure, in the strict sense of that word. Rain-water, even when collected before it has reached the earth or washed the roofs of houses, is nevertheless impure; for it contains particles of dust swept down in its descent. I leave out of the account muddy water that owes its turbid condition to pelting rain or driving storm, also sea-water with its inevitable mixture of salt and its repulsiveness to the taste. Suffice it to say that all water in its natural state and containing however slight an admixture of foreign substances is unfit for certain manufacturing purposes, for example certain delicate operations in dyeing. Very often water may be most excellent for drinking, exactly suited to domestic uses, so irreproachably clear that the sharpest eye can detect in it no alien substance whatever; and yet for such purposes as those I have indicated it may be worthless. To give water the purity required in certain of the arts it is customary to distil it, not with an apparatus of glass, such as we use for a simple experiment like the one we have just been considering, but with a more substantial, more capacious outfit. The water to be purified is poured into a copper boiler, or alembic, or cucurbit, as it is variously called, which is sometimes provided with a hot-water jacket, and sometimes is placed directly over the fire. The steam ascends to a sort of dome, or head, surmounting the boiler, and thence by a long neck, called the rostrum or beak, it reaches a metal tube coiled in a spiral and hence known as the worm. This latter is immersed in cold water contained in what is called the refrigerator. In circulating through the worm the vapor becomes chilled and is condensed into water, which runs out at the lower and free end of the worm, the latter passing through the side of the refrigerator at its base.

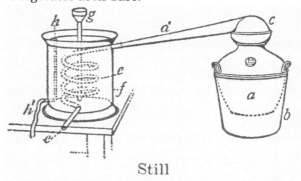

Still

"It is plain that the water in the refrigerator must gradually become heated by the steam circulating in the worm, and thus be rendered unfit for condensing purposes. Hence it must be renewed from moment to moment, and this renewal is in fact made to go on continuously. Fresh water is run into the funnel shown in the picture as reaching

a, alembic; *b*, hot-water jacket; *c*, head; *d*, rostrum or beak; *e*, worm; *f*, refrigerator; *h*, *h'*, tube for conveying away the warm upper stratum of water heated by the worm.

to the bottom of the refrigerator, while the warm water, being lighter than the cold, rises to the top and runs out through the tube also shown. There is thus a constant renewal of cold water at the bottom of the refrigerator, with an equal outflow of warm water at the top. At the end of the operation there is found in the bottom of the boiler a muddy paste representing the impurities contained in the water subjected to distillation.

"Nothing is more disagreeable than to gulp down by accident a mouthful of sea-water. Nor is this water any better for washing linen than for drinking, since it will not dissolve soap; and it is equally unsuited for purposes of cooking. But by being distilled the water of the ocean, so unfit for our use in its natural state, becomes purified. Great ocean steamers are provided with distilling apparatus in which sea-water is freed from its salt, and the resulting liquid differs not a particle from that obtained by distilling fresh water. It is suitable for cooking and washing, but not the best kind of water to drink, because it holds in solution no air, a little of which is needed in all drinking-water. But it can be made to absorb the lacking ingredient by being shaken up in contact with the air."

Chapter Forty-Four - Water

"THE water that we use every day is hardly ever pure. However clear it may be, it always contains certain foreign substances in solution, as is proved by the slight coating of earthy matter that forms little by little on the inside of our water-bottles, tarnishing the glass and lessening its transparency."

"That earthy coating is very hard to wash off," remarked Marie. "I remember one day I tried and tried to get it off with water alone, but it seemed to have become a part of the glass itself."

"Yes, that coating sticks so fast just because it is of stony matter, of veritable stone such as the mason uses for building our houses. It is not at all surprising, therefore, if mere washing with water fails to remove it. To make it let go its hold it should be dissolved in an acid, vinegar for example, or lemon juice. Pour a little vinegar into a carafe and shake it up until it has wet all the clouded part of the glass; you will see the stony coating dissolve, creating a little foam as it does so. When the acid has done its work, wash it off with water, and you will find that all the foreign matter comes away with it, leaving the glass once more as clear and transparent as ever."

"Then even the clearest water," Jules observed, "water in which the eye can detect nothing, absolutely nothing, nevertheless contains dissolved stone, just as sweetened water contains sugar invisible to the eye; and when we drink a glass of water we drink with it a little of this stony matter. Who would ever suspect it!"

"It is very fortunate, my dear boy, that we do thus drink a little of this dissolved stone. Our bodies, in order to grow and become strong, require a certain proportion of stony matter for the formation of our bones, which are to us what its solid framework is to a house. This needed matter we cannot by any possibility create by ourselves; we must get it from our food and drink. Water, for its part, furnishes a good share, and if it did not contain the required mineral matter we should remain puny and ailing, being unable to attain our natural size."

"Is there much of this dissolved stone in the water we drink?" asked Emile.

"To be fit for drinking, water must contain a little, for the reason I have just explained; but when it contains too much it is hard to digest and burdens the stomach. The right proportion is from one to two decigrams for a liter of water; or, in other words, about as much as you would take up between your thumb and forefinger. Any considerable excess makes the water heavy, as we say, because it weighs on the stomach.

"Certain waters are so rich in dissolved stony matter that they quickly encrust anything they touch. Such is the water of the celebrated spring of Saint

Allyre at Clermont-Ferrand. It is made to fall upon a heap of tree branches which break up the water and divide it into spray. This fine shower is allowed to fall on objects that it is desired to coat with an incrustation of stone — on birds' nests, baskets of fruit, bouquets of flowers, and foliage. A layer of stony matter is soon deposited by this mineralized dew, and the birds' nest, the basket of fruit, the bouquet are turned to stone, or, more exactly, they are overlaid with a coating of stone, so that one would say a sculptor's chisel had deftly cut these objects out of marble. Such water, needless to say, is unfit for drinking."

"I should think so!" cried Claire. "It would pave the stomach with marble, which would not be very good for the digestion."

"Never does the water such as we use," Uncle Paul continued, "have anything like that superabundance of stony matter, though it often does contain enough to cause difficulty in certain domestic operations, especially laundry-work. You must have noticed how the water in which clothes are washed with soap always turns more or less white; perhaps you have even observed that little flakes or clots of whitish matter are formed in the water and float about in it."

"Yes, I know what you mean," Marie hastened to reply; "and when there are too many of those white clots it is hard to get any suds; the soap is just wasted."

"Well, now you will know that the white tinge and the floating particles are caused by the presence of dissolved stony substances. Perfectly pure water, distilled water, takes up soap without losing its clearness, or with very little loss; it does not turn white, it does not form flakes. To convince yourselves of this, try a little rain-water some day for washing out a piece of linen; for rain-water is almost as pure as distilled water. You will see how easy it is and how the soap does its work without waste. There will be no white particles left in the water, though there will be plenty of lather, and no whitish tinge to the water under the foam such as you commonly see in wash-tubs.

"When water turns very white under the action of soap and shows abundant flakes, it is a sure sign of too much stony matter in the water. Laundry-work then becomes difficult and soap gives trouble about dissolving, dissipating itself in tiny clots without acting on the soiled linen. Such water is also bad for drinking, overburdening the stomach with its excess of mineral matter. The water found in regions rich in limestone is liable to this objection."

"I can see well enough," said Emile, "that a little stony substance in the water must be a good thing for us, and I also see how troublesome too much must be. The stomach would soon get tired of digesting stone."

"Finally," his uncle continued, "hard water like that is unfit for certain culinary purposes, particularly cooking vegetables such as green peas, and chick-peas — the last named especially. The mineral matter in the water becomes incorporated with the vegetables and then, no matter how long you boil them, they will not become soft."

119

"Yes," said Marie, "I know how chick-peas act sometimes: after hours and hours of cooking they are just as hard as at first and will bound like marbles if you throw them on the floor. What prevents their softening, you say, is the stony matter dissolved in the water."

"That, and nothing else. Now, since all water contains more or less of this, we are often troubled about cooking the vegetables I have named. But there is a simple remedy that I recommend in all such cases: drop a little pinch of potash into the water, and the most obstinate beans or peas will cook to perfection, even the chick-pea itself softening to a mush."

"Without getting any bad taste?" asked Marie.

"Without getting any bad taste or anything else that need be feared, on condition, however, that the potash be used very sparingly — just a pinch and no more.

"But there is another way to use it that is more readily at our command. Since potash is obtained from wood-ashes it is plain that wood-ashes can here play the part of potash. In a small piece of cotton cloth folded two or three times tie up a thimbleful of clean ashes, and drop this into the pot with your vegetables. The potash in the ashes will dissolve and permeate the water, while the earthy matter will be left in the cloth, which is to be taken out when the vegetables are done. By this means, however hard the water, you will get the better of the most refractory peas and beans."

"Uncle Paul is always finding some new use for wood-ashes," remarked Claire; "and now we see that they will soften even the hardest of chick-peas."

Chapter Forty-Five - Water (continued)

"WATER may be clear, colorless, fresh, agreeable to the taste, excellent for washing and cooking, and nevertheless, with all these admirable qualities, dangerous to drink. This danger arises from microbes in the water, though nothing betrays their presence, neither smell nor taste, nor any lack of clearness, nor the least impairment of the water for household uses. Since certain kinds of these infinitesimal organisms cause serious maladies, we imperil our health by taking them into our bodies in our drink. Water, therefore, to be good to drink, should contain no microbes. One well may furnish water of irreproachable purity, while another may be more or less infected with microbes and hence pernicious and dangerous despite all appearances to the contrary.

"Where, then, are we to look for perfectly pure water, water that we may drink without thought of danger? It is furnished only by springs. Let us dwell for a moment on the origin of springs, and we shall then understand why spring-water is pure. Rain, melting snow, the dampness of night fogs, soak into the ground, especially on mountain slopes; and the water thus absorbed over large expanses of surface sinks to a great depth, collects in little underground streamlets, makes its way through opposing soil and sand, also through the cracks in rocks, and comes to the surface again in some distant valley, welling up through a fissure and producing a spring.

"From its starting-point to its destination the water thus passes through a sort of filter of enormous thickness, kilometers thick in fact, and at a sufficient depth to be free from surface defilement. By passing through successive beds of clay, marl, sand, crumbling rock, and porous stone, the water gradually rids itself of its impurities and leaves them behind, so that on reappearing above ground it no longer contains any corpuscles even of the microscopic minuteness of microbe germs. Spring-water is pure by virtue of the thorough filtering it has undergone, a filtering such as no means at our disposal could begin to achieve.

"Can we say as much for the purity of river water and brook-water? Far from it. These streams, especially in the neighborhood of large cities, receive frightful quantities of foul matter. Into them empty sewers charged with the refuse from streets and dwellings; in their waters are washed the garments we have soiled and the foul linen that has served as bandages for sores; their channels are choked with all sorts of decaying matter from many factories. It is therefore evident that river and brook, however clear their water, cannot furnish us with a drink that is free from suspicion. Microbes abound, and those of cholera, for example, may be among them, from the person of some victim to that disease or from the linen used in his treatment.

"Not even a country brook is void of offense. Rain-water washes the tilled fields, soaks through the manure spread as fertilizer, and carries to the stream the harmful germs that breed in all decay.

"Well-water, besides being not always sufficiently aerated, is likewise subject to defilement. In the first place, owing to its slight depth, a well becomes charged with water from the upper layers of the soil, a filter not thick enough to arrest injurious germs. In the second place, wells in towns are dug in ground that has become defiled to a considerable depth by the prolonged sojourn of man. Not far away are drains and sewers and other repositories of filth, from which it is very difficult to safeguard the wells.

"In the country the danger is less, provided the well be covered so as not to admit any dead leaves or the dust raised by the wind; and provided especially that the well be at a distance from all stables, dung-heaps, deposits of compost" and other sources of infection through infiltration.

"Mere taste and appearance make us reject for drinking-purposes all water that repels by its odor, its taste, or its lack of clearness. But this is not enough. It is now established beyond doubt that certain diseases, especially typhoid fever and cholera, are propagated by water containing their microbes. Suspicious as we must at all times be of river-water and well-water, in periods of epidemic we should exclude them entirely from our use and have recourse to spring-water alone.

"But not every one can obtain spring-water. What shall be done in such cases? The answer is simple. We have seen that the temperature of boiling

121

water kills all living creatures. River-water or well-water can accordingly be rendered quite fit for all our uses on the express condition that it be first boiled. Freed of its noxious germs by heat, it is thenceforth harmless.

"Summing up these points in a couple of precepts of prime importance to our health, we may say: Keep all wells and springs free from filth, and when cholera is prevalent use no well-water or any water from river or brook without first boiling it."

Chapter Forty-Six - Vinegar

"YOU will be surprised to hear," said Uncle Paul, "that any sweetened substance will generate alcohol by a remarkable chemical change called fermentation, and that alchohol in its turn changes into vinegar. As sugar is the origin of alcohol, it is sugar, in reality, that makes vinegar. Here we see something generating its opposite, sweet giving birth to sour."

"The same thing happens," Marie observed, "with milk or with a slice of melon: they both sooner or later lose their sweet taste and turn sour.

"Those are two good examples of substances which, at first sweet, turn sour as soon as decomposition sets in; but vinegar such as is used in cooking goes through a little different process; for it comes not directly from sugar but from alcohol. All alcoholic liquids are good for making vinegar; nevertheless wine makes the best and most highly valued. The very word vinegar shows you how the thing itself is made, 'vinegar' meaning nothing more nor less than 'sour wine' — *vin aigre*."

"Why, so it does!" Claire exclaimed. "I hadn't noticed it before. The two words fit together just right; not a letter too many, and not a letter too few."

"In wine," Uncle Paul resumed, "it is the alcohol, and the alcohol alone, that turns sour. That is to say, you can't make good vinegar without good wine. The more generous the wine, or, in other words, the richer it is in alcohol, the stronger the vinegar. People often make a mistake on that point: they think that poor wine, the final drippings from the winepress, the rinsings of bottles and casks, will in course of time take on sufficient sourness. A great mistake. Such watery stuff cannot possibly yield what it does not possess. As soon as the small proportion of alcohol it contains has turned to vinegar, that is the end of it; no matter how long you wait, there will be no increase of sourness. The rule has no exceptions: to obtain good vinegar use good wine, wine rich in alcohol."

"But you haven't told us yet," said Jules, "what must be done to change the wine into vinegar."

"That takes care of itself. Leave on the kitchen sideboard an uncorked bottle of wine, not quite full, and in a few days, especially in summer, the wine will turn to vinegar. On the express condition of its being exposed to the air, wine will turn sour of itself, and all the quicker when a warm temperature hastens the process of decomposition in the alcohol. That shows you at once the care necessary for keeping table wines and preventing their turning sour. If in bottles or demijohns, they must be tightly corked with good stoppers, since otherwise air will get in and the wine will be in danger of souring. As cork is always more or less porous, the top is covered with sealing-wax when the wine is to be kept a long time; in a word, the bottles are sealed."

"Then it's just to keep out the air, "said Emile, "that they seal the bottles with red, green, black, or any other colored sealing-wax?"

"Merely for that reason. Without this precaution air might gradually get into the bottle, and when it was uncorked, instead of excellent old wine, you would have nothing but vinegar. You see, if you wish your wine to keep well, you must, above all, guard it from the air. A partly filled demijohn or cask, opened every day to draw out wine and then carefully recorked, soon goes sour, especially in summer. If the wine is not likely to be all used up for some time, the contents should be bottled and carefully corked. In that way the wine is in contact with the air only one bottle at a time, as it is called for, and so cannot turn sour provided it has been properly corked.

"Let us, then, accept it as a rule that if wine is not to turn sour it must come in contact with the air as little as possible. If, on the contrary, we wish to change it into vinegar we leave it exposed to the air in uncorked or imperfectly corked vessels. Little by little, through the long-continued action of air, its alcohol will turn sour. That is what happens to the remnants of wine left in the bottom of bottles and forgotten.

"Of all the seasonings used with our food, vinegar, next to salt, is the most prized. With its cool, tart flavor and agreeable odor it gives a relish to dishes that without it would be too insipid. Its use is not only a matter of taste, but also of hygiene, for taken in moderation it stimulates the work of the stomach and makes the digestion of food easier. Combined with oil it is an indispensable seasoning for salad. Without it this raw food would hardly be acceptable to the stomach."

"That is one of my favorite dishes," Jules declared, "especially when it is made of spring lettuce; the vinegar makes it taste so good, pricking the tongue just enough and not too much."

"Vinegar is also used in the preparation of certain well-known condiments — capers, for example."

"Oh, how I like them!" cried Emile, "Those capers they sometimes put into stews. Where do they come from?"

"I will tell you. In the extreme south of France, near the Mediterranean, there is cultivated a shrub called the caper-bush. Its favorite haunts are rocky slopes and the fissures in old walls and rocks much Caper-Bush exposed to the sun. Its branches are long and slender, armed with stout thorns. These branches bend over in a graceful green mass, and against the darker background of foliage are set off numerous large and sweet-smelling pink blossoms resembling those of the jasmine. Well, these blossoms, before they open, are capers. As little buds they are gathered every morning, one by one,

123

and pickled in vinegar of good quality. That is all that is done to them. So when Emile smacks his lips over the caper sauce, he is eating nothing more nor less than so many flower buds."

"I shall like them all the better for knowing they are flowers," the boy declared.

"In like manner gherkins are pickled in vinegar. They grow on a vine much like the pumpkin-vine. Similar treatment, too, is given to pimentos, sometimes called allspice on account of their spicy taste, which becomes unbearably strong when the fruit is ripe and coral-red. I will remind you that all pickling with vinegar should be done in vessels not glazed on the inside with lead. I have already told you that ordinary pottery is glazed with a preparation that contains lead. Strong vinegar might in the long run dissolve this glaze and thus acquire harmful quali-

Caper-Bush

ties. Keep your capers, pimentos, and gherkins in glass vessels, or at least in pots that are not glazed inside.

"In conclusion I will tell you that vinegar has the property of making meat tender. To insure tenderness in a piece of beef it is sprinkled several days in advance with a little vinegar to which have been added salt, pepper, garlic, onions, and other seasoning, according to the taste of each person. In this mixture, however many of these ingredients there may be, vinegar plays the chief part. This process is called sousing the meat."

It is a little strange that although excellent cider is produced in France, especially in Normandy, it seems not to be used for making vinegar. — *Translator's Note.*

Chapter Forty-Seven - The Grist-Mill

"THE supply of flour in the house was getting low. Accordingly a trip to the mill was undertaken. The donkey bore a sack of wheat across his back, and as he made his way along with short steps the patient animal cropped here and there from the hedges a mouthful of thistle blossoms, a rare dainty to his palate. The children ran this way and that, picking wild flowers or chasing butterflies, and returning now and then to Uncle Paul with some treasure from Mother Nature's store to exhibit to him.

Before long the sound of falling water and the *tick-tack* of the mill, from its half-concealment amid the foliage of willows and poplars, fell on the ear. Ducks were returning in single file from their bath, some of them as white as snow, with feet and beak orange-yellow, others with head of bright green, wings emblazoned with a splendid blue spot in the middle, and tail surmounted by a little curled feather. Ducklings were picking up, here and there, with noisy demonstrations, some scattered grains of wheat along the way. Startled by the approach of the donkey, a flock of geese extended their necks and uttered a raucous cry of trumpet-like resonance, to which the donkey replied after his fashion. And so the journey's end was reached.

After settling certain business details with the miller, Uncle Paul proceeded to show his young charges over the mill and explain its working.

"The water from the stream," said he, "is stored up in a large reservoir by means of a dam, and is let out by opening a sluice-gate. Through this gate it rushes with great force and strikes against the floats of a water-wheel set directly in its path, causing the wheel to turn amid a shower of spray and foam, and thus setting in motion the machinery of the mill.

"They grind the grain and reduce it to flour there are two large millstones, very hard and arranged one over the other, their flat surfaces very near together but not touching. These surfaces are rough, the better to seize and grind the grain between the two disks, the lower one of which is motionless, while the upper revolves rapidly under the impulse of the waterwheel. Both are enclosed in a round wooden case which keeps the flour from scattering. The upper stone has a hole in the middle, through which, little by little, falls the wheat contained in a sort of large wooden funnel called the hopper. Into this the miller empties the grain-sack. As fast as it comes between the two millstones the grain is caught by the irregularities in the surface of the revolving stone and crushed against those in the stationary one. The resulting powder, the coarse flour, is driven by centrifugal force to the edges of the millstones, and finally escapes in a continual stream through an opening in the front of the wooden case.

"But on thus issuing from between the millstones the flour is not yet ready for making into fine white bread; the bran from the husk of the grain must first be removed. This is done by means of a sieve made of silk, which receives the coarse meal, lets the fine flour pass through its meshes, and retains the bran. The mill has now done its part. To it came sacks of wheat which, unground, could not serve as food; from it go sacks of flour to furnish bread for daily use. Water-power, harnessed to the machinery of the mill, accomplishes this important transformation, turning the millstone to grind the grain and revolving the sieve to separate the bran from the flour. The miller's task is confined to watching the wheels as they turn and feeding the hopper with grain.

"What weariness and waste of time if we had not the aid of machinery and were compelled to grind our grain by sheer strength of arm! You must know that in ancient times, for lack of better knowledge, people had to crush their wheat between two stones after first parching it slightly over a fire. The coarse meal thus obtained was boiled in water to a sort of porridge and then eaten without further preparation, as I have explained in one of our former talks." [1]

"And what did they do for bread!" asked Jules.

"Such a thing as bread had never been even dreamt of at that time. Wheat was eaten only in the form of porridge, a sort of thick glue, its insipid taste somewhat relieved by parching the grain a little to begin with, as I have already explained. Later the plan was hit upon of making a dough of flour and water and then baking this on the hot hearthstone, thus producing some very inferior pancakes, as thick as your finger, stodgy and hard, and mixed with ashes and charcoal. These were better than porridge, but far inferior to the poorest bread of today. By repeated trials, however, our ancestors at last succeeded in producing bread like that on our tables at the present time.

"Next they had to devise means for grinding wheat in considerable quantities, but their invention fell far short of our modern grist-mill. A hollowed-out stone was used as a mortar, and into this was fitted an unwieldy pestle which was operated by a bar pushed by wretched slaves, under the compulsion of a cruel driver armed with a rawhide. Thus slowly and painfully were a few handfuls of flour produced in the time taken by one of our grist-mills to grind a barrelful."

[1] See "Field, Forest, and Farm."

Chapter Forty-Eight - Bread

MOTHER AMBROISINE was making bread. Standing in front of the kneading-trough, her cheeks glowing with the exercise, her sleeves rolled up to her elbows, alternately she thrust her closed fists into the mass of dough, which yielded with a dull *flic-flac,* then, lifting in both hands ponderous sheets of it, she let them fall back heavily into the trough. Marie, standing on a little stool so as to bring herself on a level with the trough, lent what aid she could in this rather arduous task. After sufficient kneading the dough was cut into a number of pieces, each of them destined to become a loaf of bread, and these were put into little straw baskets and covered with woolen cloths so that a gentle heat might finish the work already begun.

Before the loaves had risen enough for baking. Uncle Paul gathered the children around him and told them some interesting facts about bread-making.

"If the flour were merely mixed with water," he began, "and the dough were then put into the oven in that condition, the result would be nothing but a dense, heavy loaf, a sort of hard pancake, repulsive to the stomach on account of its indigestibility. Bread, to be easily digested, must be full of countless little holes, like a sponge; it must have those myriad eyes that help to crumble the loaf and aid the final process of subdivision which it is the stomach's part to perform. A prolonged diet of bread made simply of flour and water is so far from appetizing that it was imposed as a penance by the Israelites in one of their sacred festivals. When they set out from Egypt, led by Moses, they had not time in their hasty departure to prepare bread in the ordinary manner, and so they cooked in the ashes cakes called unleavened bread, that is to say bread without any leaven or yeast. In commemoration of this event the Jews of our own day eat unleavened bread at the celebration of their Feast of the Passover. This bread consists of thin cakes, compact and hard, of which a few mouthfuls are not unpleasant, but if one eats nothing else for several days the stomach is left unsatisfied. It is by a fermentation similar to that of must in wine-making that flour is made into bread, real bread, the most highly prized of all foods and the one we never tire of.

"Flour contains, as I have already told you, [1] first starch and gluten, and then a small quantity of sugar as proved by the slightly sweet taste of a pinch of flour on the tongue. Now this tiny proportion of sugar is precisely the substance that causes fermentation in dough; that is to say, it decomposes and forms alcohol and carbonic acid gas, as in the making of wine."

"Bread and wine, then," said Marie, "are something alike in the way they are made."

"It is more than a likeness: it is a perfect sameness so far as concerns the decomposition of sugar into carbonic acid gas and alcohol; and it is, finally, a sameness in respect to fermentation. The dough of which bread is made ferments just as does the must that is to become wine.

"It remains for us to see how this fermentation started. Nothing is simpler: one has only to mix with the new dough a little of the old dough kept over from the last bread-making and called leaven. or rising. This old dough has the peculiarity of making sugar ferment, of decomposing it into carbonic acid gas and alcohol. The noun 'rising' comes from the verb 'rise,' because by virtue of the rising mixed with it the dough rises, swollen by the carbonic acid gas that is generated.

"Rising, as I have just said, is fermented dough left over from the last bread-making. It is lukewarm to the touch on account of the process of decomposition going on within."

"Grape juice gets warm, too, when it ferments," remarked Claire.

"Furthermore," Uncle Paul went on, "rising is much swollen and very elastic on account of the gas imprisoned within its glutinous mass; and it has a pungent, winy odor from the alcohol formed by the decomposition of the sugar it contains. Such, then, is the indispensable ingredient called for, even though in only a small quantity, in order to make fresh dough capable of becoming bread such as we all like so much. To please the palate, salt is also added, but it performs no other office.

"The kneading done, what comes next? I will tell you. Acted upon by the rising that has been evenly mixed with the entire mass of dough, the sugar

127

therein contained decomposes. The carbonic acid gas thus generated remains imprisoned, the gluten merely dilating under its pressure and forming a spongy mass of membranous tissue packed with innumerable tiny closed cavities. Thus the dough rises, swells, and becomes riddled with holes like a sponge. Baking increases this porosity, for the gas, finding itself restrained by elastic glutinous walls, expands still more with the heat and makes the already existing cavities larger. To its highly nutritive quality gluten adds another: by retaining the carbonic acid gas within its multitudinous cavities of all sizes it makes the bread very porous, very light, and consequently easy to digest. Hence it is that a flour poor in gluten, such as rye meal, makes only a compact bread, heavy for weak stomachs; and hence also a flour containing little or no gluten, such as the flour made from rice, from chestnuts, from potatoes, is totally unfit for making bread.

"For perfect fermentation a warm temperature is necessary. That explains why woolen coverings are put over the little baskets of dough to keep in the heat and shut out the cold. If you raise one of these covers and put your hand on the dough you will find it lukewarm and plump. It is warmed by fermentation and swollen by carbonic acid gas."

"Yes, it really is warm," Claire announced, after feeling of one of the unbaked loaves; "and when I press it, it goes in like a rubber ball and then swells out as soon as I take my hand away."

Toward evening the bread came out of the oven all golden-brown on the crust, and filled the house with a sweet smell. The children thought it tasted better than ever, now that they knew how it was made.

[1] See "Field, Forest, and Farm,"

Chapter Forty-Nine - Other Wheat Products

BESIDES the best of bread, we are indebted to wheat for macaroni, vermicelli, and other similar products manufactured chiefly in Italy."

"Vermicelli is in the shape of long worms," observed Jules.

"Yes, it is precisely from its resemblance to a mass of long worms that it gets its name of vermicelli, which means 'little worms.'

"And macaroni," Marie put in, "which makes just as good a dish, seasoned with cheese, is in the shape of long, slender tubes."

"Besides these two," Claire hastened to add, "there are any number of other shapes — stars, circles, ovals, hearts. I have even seen the letters of the alphabet; the first time I noticed them I was surprised to find every one of the letters in a single spoonful of soup. It must take a long time to cut the dough into all those little pieces, all so beautifully shaped."

"No time at all, my dear," Uncle Paul assured her. "In a twinkling, by means of machinery, there is turned out any quantity of those beautiful shapes you admire so much. Let us talk a little about how they are made.

"As you now know, the most nutritive part of wheat is the gluten, which may be compared, in respect to nourishment, with meat itself. Therefore in these Italian products wheats having the highest percentage of gluten are used. They come from hot countries, notably from Sicily, Africa, and Asia. In making the dough very little water is used, in order to get a firm dough; and to improve its taste and color it is customary to add a little salt and saffron. This dough is put into a metal case, the bottom of which is pierced with a number of holes, some round, some ring-shaped, some representing stars, some formed like hearts, flowers, letters of the alphabet, etc. — in fact, any shape desired. The dough is pressed through these openings and thus made to take the various forms I have mentioned. If the bottom is merely pierced with small round holes, the dough will come out of the case in long round threads as vermicelli."

"Why, that's as easy as saying good morning!" cried Claire.

"If the bottom is pierced with ring-shaped openings, the result will be of a larger size and known as macaroni."

"But I see one difficulty. Uncle Paul," objected Marie. "The openings in the bottom cannot be perfect rings, for then the piece in the middle would have nothing to hold it."

"Very true. Accordingly the ring-like openings are not complete, and the macaroni comes through split all down its length. But the two edges of the fresh dough unite as soon as the opening is passed, stick together, and the wall of the tube is without a break."

"Lastly, if the openings are in the shape of narrow slits, what comes through will take the form of thin ribbons or thongs."

"With openings shaped like stars," said Claire, "the dough pressed through will be in long grooved strings; but there won't be any stars such as we have in soup."

"To obtain these stars and other similar products, there is placed, a little to one side and under the case, a broad circular blade having a keen edge and revolving rapidly. At very short intervals it severs the strings of dough as they come out of their molds. Each of these segments is a star, heart, oval, crescent, flower, or letter, according to the shape of the orifice that molds the string."

"Now I understand," Claire rejoined. "A big knife goes backward and forward at the mouth of the mold, and there falls from the bottom of the case a shower of stars coming from the strings of dough which the knife cuts off very short. That is the way I get those little rounds of carrot that have such a beautiful yellow color they remind me of gold coins. You told us that the dough for making all these things is colored with a little saffron. I don't know what that is."

"Saffron is a plant cultivated in some of our departments, especially around Angouleme and Nemours. It yields a magnificent orange-yellow color, contained in three long and slender threads found in the very center of the flower. The flowers are gathered as fast as they open, but only these three thin threads are kept. The harvest is limited to this small part of the flower, so I leave you to imagine how many blossoms it takes to furnish a little of this coloring matter. [1] These tiny threads, dried in the sun and reduced to powder, constitute the coloring-matter with which a beautiful yellow tint is given to dough for macaroni and similar products, as also to cakes, butter, and cream. Furthermore, saffron is sometimes used, in very small quantities, as a seasoning."

"Besides all these things you have been telling us about," said Marie, "I've often have groats in our soup, and groats I take to be nothing but wheat grains without the bran."

Saffron

"Since wheat, the richest in gluten of all the cereals, furnishes us macaroni and vermicelli and other similar food products, it is clear that it can itself serve as food without being first reduced to flour and then made into dough. We have simply to remove its outside covering by running it between two millstones only the thickness of a grain apart. The two stones, not being close enough together to crush the wheat and reduce it to flour, merely take away the coarse outside skin, that is to say the bran. The result of this operation is called groats, which we may regard as a sort of natural macaroni or vermicelli obtained at little cost with a few turns of the mill-wheel.

"Oats and barley also furnish groats. What is called pearl barley is barley groats rounded under the millstone in the form of little globules. Lastly, semolina is groats reduced to very small grains. In a word, it is wheat half ground and hence in particles coarser than ordinary flour. Another kind of semolina is made of dough such as is used for vermicelli: the dough is chopped up into a sort of fine sand instead of being molded in long strings."

[1] A single ounce requires the product of more than four thousand blossoms. — *Translator.*

Chapter Fifty - Strange Uses of Starch

AFTER explaining the important part played by starch in plant life Uncle Paul took occasion to name some of the less familiar sources of this substance and to describe how it is used for purposes very different from that so well known to every housewife and every laundry-maid.

"All starch," said he, "whether it be derived from one plant or another, from a seed or from a root, is readily convertible into sugar either by the natural processes going on in vegetation or by artificial processes employed by man.

The simplest expedient is the application of heat, a factor entering into the preparation of farinaceous foods. Let me illustrate by a few examples.

"A potato in its raw state is uneatable. Cooked in boiling water or roasted in the ashes it is excellent. What then has happened to it? Heat has turned a part of the starch into sugar, and the tuber has become a mass of farinaceous dough, slightly sweetened. We can say about the same of the chestnut. Raw it is not good for much, though at a pinch it can be eaten; cooked, it deserves all the praise we give it, and I am sure you will back me up in this assertion. Here again we have a transformation of starch to sugar by the action of heat. Beans and peas, hard as bullets when dry, and far from pleasing to the palate, are unmistakably sweetened as soon as boiling water has worked upon their starch. Our farinaceous foods of sundry sorts behave in similar manner."

"Then do we make sugar," asked Claire, "whenever we boil a pot of potatoes or chestnuts or beans? I didn't know we were so clever; but I for my part shall not put on any airs, as it isn't very hard to make a pot boil."

"Man's ingenuity has devised a more effective means than heat alone for converting starch into sugar. The starch is boiled in water and, while it is boiling, there is added a small quantity of a powerful liquid called oil of vitriol or sulphuric acid. This causes the starch to turn to syrup, after which the oil of vitriol is of course removed. It has done its office. The substance thus obtained is soft, sticky, and nearly as sweet as honey; it is called starch-sugar, or glucose, and is much used by confectioners. As I have already told you, the sugarplums and other sweets that you buy at the candy shop are in most instances the product of this ingenious process of turning starch into sugar. And so you see the humble potato furnishes you with something besides the modest dish you find every day on the table.

"But that is not the whole story. Glucose, obtained as I have described, is exactly the same as the sweet part of ripe grapes. With potatoes, water, and a few drops of sulphuric acid there is artificially produced, in enormous boilers, the same sweet substance that nature manufactures by the action of the sun's rays on the full-grown grape. Now, since grape-sugar turns to alcohol by fermenting, starch sugar ought to undergo a like transformation. As a matter of fact, in northern countries, where the climate is too cold for the vine, alcoholic liquors are made from starch that has first been changed to sugar. Such liquors bear the generic name of potato-brandy, though all seeds and roots rich in starch may be used in the same way as the potato."

"Let us now drop the subject of potato-brandy, which I have briefly touched upon to satisfy your curiosity, and return to matters of household economy. There are various starchy substances that are much used in making soup, chief among them being the starch of potatoes, which furnishes a nourishing and appetizing dish of this sort and is our most important, least expensive, and most widely distributed food product of its kind. Many of the starchy preparations bearing pretentious names are really nothing but this, at least in part. Other forms of the same essential substance appear more rarely on our tables, their higher price causing them to be reserved for dyspeptics and convalescents. Let us consider for a moment the chief of these.

"In South America there is cultivated a large farinaceous root called manioc, which in its natural state is a deadly poison to man, but which nevertheless furnishes material for excellent bread. First the root is reduced to pulp with a grater, after which the juice is squeezed out, and with the juice goes the poison, leaving a harmless substance rich in starch and serving as the principal article of food for the poor in a country too hot for raising wheat. This farinaceous substance is sold with us under the name of tapioca. A spoonful of tapioca is transformed by the action of boiling water into a rich jelly of exquisite fineness.

"The woods and meadows of our own latitude abound in certain plants known as orchids, remarkable for their oddly shaped flowers and for the two small tubers of the size of pigeons' eggs in the midst of the fine roots of the plant. These tubers contain starch. They are gathered in eastern countries, and flour made from them comes to us under the name of salep or salop. Prepared with hot water, it furnishes a gummy jelly suitable for the use of invalids.

"Palm-trees grow only in a hot climate. The trunk of a palm is a graceful column, without branches, of lofty height, tapering but little from bottom to top, and crowned with an enormous tuft of large leaves. One of these trees, the sago-palm, has the heart of the trunk filled with a farinaceous pith which is removed after the tree is cut down. From this pith is obtained a starchy substance known as sago and differing only slightly from potato starch.

"These strange forms of starch, which excite our curiosity but are of no great use to us, must not make us forget the farinaceous matter furnished by our own leguminous plants, lentils, beans, and peas. You know the excellent thick soup we make of dried peas, and you doubtless also know how disagreeable are the hulls of this vegetable, tough as parchment and without taste or nourishment."

Flowering Orchid

"Yes," replied Emile, "if it were not for those horrid hulls, dried peas wouldn't be at all bad."

"The hulls, however, are got rid of by pouring the soup into a colander, which retains the objectionable part and lets through the pure pulp. But in the process a certain quantity of nutritive matter mixed with the hulls is lost.

"Invention and experiment have done away with this loss. The peas are steeped a few minutes in boiling water to burst the hulls, after which the peas are dried in an oven and then made to pass between two millstones suf-

132

ficiently far apart to remove the hulls without touching their contents. Thus freed of their tough exterior, the peas are ground to powder, which goes under the name of pea starch. In similar manner are obtained bean starch and lentil starch. All these preparations are used for making soup, and all have the qualities, without the defects, of the vegetables from which they are derived; that is, they are freed from the disagreeable hulls that fatigue the stomach to no purpose."

Chapter Fifty-One - Rice

ONE day Emile showed a sulky face because, when he went to Mother Ambroisine for something to eat between meals, she gave him only a slice of bread without butter or honey or anything else on it to make it taste good. But Marie reproved him, saying there were plenty of people in the world that would be glad enough to get a slice of dry bread and would even consider it a royal feast.

"For it isn't every one can have bread when he wants it," she continued. "There are countries where the people have never even seen such a thing. Isn't that so. Uncle Paul?"

"It is only too true," was his reply. "You already know from my talks with you that not even in our own favored land can all the people have white bread on the table. In many homes rye and barley serve as very inferior substitutes for wheat; and what is true of this country is even more notably the case throughout large sections of the world as a whole."

"But what do people eat if they can't get bread of any kind!" asked Claire.

"Sometimes one thing, sometimes another. There are a number of cereals, some of them quite unfamiliar to us, that afford nourishment, though furnishing nothing like our light and fragrant white bread with its crisp crust and sponge-like interior. Asia has rice, Africa millet, and America maize, or Indian com. In China and India the people have hardly any food but rice cooked in water with a little salt. In fact, half the world lives on virtually nothing else."

"Rice, then, takes the place of bread with those people, doesn't it?" asked Claire.

"Yes, it may be said to take the place of our bread when they have anything to go with it; but not infrequently the whole meal consists of rice."

"With nothing else, at all?" asked Emile incredulously.

"With nothing else of any description," his uncle assured him, "from year's end to year's end."

"Then they must be an uncommonly frugal sort of people."

"Yes; but the warmth of the climate makes this light diet sufficient, whereas in our latitude, with its colder temperature, we should die of consumption if limited to such fare."

133

"Is this rice that takes the place of bread in China and India really the same as that we buy at the grocer's?" asked Claire. "We sometimes have that cooked with milk."

"Exactly the same. It is imported into this country from distant lands. What you had last week, as soft as sugar and as white as snow, may have come from the country of the Hindus, or perhaps from China. The plant producing this article of food has a stalk not unlike that of wheat: but instead of the latter's erect ear of grain it bears a graceful tuft of weak and drooping clusters of seeds. The leaves are long and narrow, like ribbons, and are rough to the touch. It is an aquatic plant, as you have learned in one of our former talks, [1] requiring a marshy soil and growing almost submerged in mud and water. Artificial irrigation is often resorted to in China to bring about the needed conditions, and when the harvest season arrives the water is drawn off and the reaper, sickle in hand, wades into the mud to garner the heavily laden tops of the rice-stalks. But it is a task far different from our cheery harvest; there is no chirping of crickets or song of lark to enliven the work, no display of corn-flowers or poppies to gladden the eye. The reaper plies his sickle with the mud and water reaching sometimes as high as his knees."

[1] See "Field, Forest, and Farm."

Chapter Fifty-Two - Chestnuts

"I HAVE explained to you what ingredients flour should contain to make it suitable for bread. It must have both gluten and starch. All flours are rich in starch, but very few possess gluten, so valuable for its highly nutritive qualities and its peculiarity of expanding in delicate membranous tissue when the dough ferments. You have not forgotten that the carbonic acid gas generated by fermentation remains imprisoned in the dough, held in confinement by the gluten, and so causes the formation of innumerable empty spaces or tiny cells which should be found in all bread worthy of the name. If gluten is lacking, these eyes also are lacking, and the dough makes nothing but a dense cake wholly unworthy of the name of bread. Well, rice and maize both furnish very white flour pleasing to the eye but deficient in one essential: it has no gluten. For that reason neither rice nor maize will make good bread, in spite of the fine appearance of their flour.

"Sometimes maize is used for making what are called corn pones, which well illustrate the difference between bread made of wheat and bread made of a flour containing no gluten. These cakes have a crisp crust that is very good to look at, but their taste does not correspond with their inviting appearance. They are but coarse, indigestible eating, and after a few mouthfuls you will be glad to desist unless you have a very strong stomach. I class the deceptively inviting corn pone in the same category with barley bread, or even below it. But, as I have explained in one of our former talks, [1] maize has its uses as a wholesome food among the farmers who raise it and whose active outdoor life enables them to digest coarse fare."

"I see more clearly every day, "said Claire, "that all those foreign grains, from Asia and from America, are far inferior to our wheat."

"I'd rather have a slice of bread, any time," declared Emile, "than all the hasty pudding or porridge or boiled rice you could offer me."

"Even without any butter on it?" his uncle asked.

"Yes, even without butter."

"I am very glad our talks are leading you to value bread at its true worth. If we were obliged to do without it now that we have become used to this incomparable food, you may well believe it would be the severest of privations.

"Your mention of hasty pudding reminds me of another well-known porridge called *polenta*, the national dish of Corsica and part of Italy. It is made of chestnut flour. Let us first say a few words about the tree that produces these delicious nuts, which you all like so much either boiled or roasted.

"The chestnut is a tree that lives to a great age and attains enormous dimensions. In our mountainous districts I have seen some with trunks four meters in circumference, and the trees must have been from three to four centuries old. One of these giants would be enough to shade my whole garden. The largest tree in the world is a chestnut growing on the slopes of Mount Etna in Sicily. It is called the chestnut of a hundred horses, because Jeanne, Queen of Aragon, visiting the volcano one day

Flowering Branch and Fruit of Chestnut Tree

and overtaken by a storm, sought refuge under it with her escort of a hundred cavaliers. Under its forest of foliage both riders and steeds found ample shelter. To encircle the giant thirty persons with outstretched arms and joined hands would not be enough; the circumference of the trunk measures in fact more than fifty meters. In its immense size it is more like a fortress or tower than the trunk of a tree. An opening large enough to permit two carriages to pass abreast tunnels its base and gives access to the cavity of the trunk, which is arranged as a dwelling for the use of those who come to gather the chestnuts; for the old Colossus, whose age runs into the centuries, still has young sap and seldom fails to bear."

"This prodigious tree must produce a mountain of chestnuts," observed Jules.

135

"I imagine one year's harvest would be enough to satisfy all of you for a long time."

"We should never see the last of them," Emile assented, "for there would be sacks and sacks of nuts — more than all the boys and girls around here could eat in a year."

Uncle Paul went on with his talk: "Chestnuts are enclosed in a husk bristling with long prickles and opening at maturity in the autumn to let the nuts fall out. There are three or four in each husk or bur. A kind of chestnut remarkable for its size and quality is known as the large French chestnut; it comes to us chiefly from the vicinity of Lyons. You must not confound the edible chestnut with that of another tree called the horse-chestnut, a tree that is often planted for ornament in parks and along streets and public promenades. Horse-chestnuts have all the appearance of the finest edible chestnuts, and are also contained in a thorny husk; but this resemblance ends with the outside, horse-chestnuts being insufferably bitter in taste and absolutely worthless as food.

"White chestnuts, or chestnuts stripped of their shells and inner skins and dried for keeping throughout the year, are obtained in the following manner. On large screens extending from end to end of a long room chestnuts are spread by the hundredweight, and under them there is lighted a fire which produces a great deal of smoke. As soon as they are well dried the nuts are put into sacks, beaten with sticks, and vigorously shaken. By this heating and shaking the shells, which have been rendered very friable by the heat and smoke, are broken into little pieces. Chestnuts prepared in this way are used boiled in water, or sometimes they are ground into flour at the mill. This flour, mixed with water and cooked over the fire for some time, gives the porridge called polenta."

"I have never tasted polenta," said Claire, "but I presume it isn't as good as fresh chestnuts roasted on the stove or simply boiled. The white, dry chestnuts you speak of are not equal to them, either."

"But they have the great advantage of keeping all the year round, while fresh chestnuts spoil in a few months."

"When chestnuts are being roasted in the hot ashes or on the stove, they sometimes burst with a loud noise and scatter the hot meat in all directions. It's funny to hear these little bombs, but I'm always afraid for my eyes. Why do chestnuts burst like that and jump off the stove?"

"Fresh chestnuts, like all undried fruit, contain a little water, or moisture. The heat of the fire turns this water into steam, which, being held captive by the tough shell and having no outlet, keeps trying to escape, until at last the overstrained shell breaks and with a rush the steam bursts out with a loud report through the rents, carrying with it torn fragments of the chestnut. To prevent these explosions, which waste the chestnuts by ripping them open so violently, and are, besides, not without danger to the eyes of those present, it is well to make an opening for the steam so that it can get out as fast as it forms, without gaining force by accumulating. This is done by making an

136

incision in the shell of the chestnut with the point of a knife, or by cutting away a small piece of the shell. Then the steam has an open door by which to escape, and the chestnuts no longer burst while they are on the fire."

[1] See "Field, Forest, and Farm."

Chapter Fifty-Three - Codfish

EMILE came to his uncle with a question. "Tell me," he began, "about the cod that has to be put to soak several days before it is eaten, in order to freshen it — isn't it a fish? Yet I don't see any head; and then it's all flat, with the bones showing on one side."

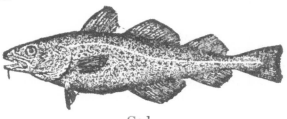

Cod

"Yes," was the reply, "the cod is a fish, and a very fine one, too, as it swims in the sea. To preserve it for keeping a long time the fishermen remove its head, which is of little value on account of its bones; then they split the body all down the stomach, throw away the entrails, and spread out the two fleshy halves, forming together a sort of slab, broad at one end and running to a point at the other. Finally, the fish thus treated are liberally salted and put to dry in the sun. So the cod reaches us all out of shape and almost unrecognizable. In its natural state it is a beautiful fish. The back and sides are bluish gray, with numerous golden-red spots like those that adorn the trout of our fresh-water streams; the stomach is silvery white; the upper jaw is prominent, while from the lower hangs a worm-shaped little barbel; and the mouth is armed with innumerable fine, pointed teeth that fringe not only the jaws but also the cavity of the mouth and throat as far down as the bottom of the gullet. And so, as you might guess from its appearance, the cod is very greedy, always in quest of food, endowed with an insatiable appetite."

"And what does it live on," asked Emile, "this greedy eater with teeth down to the bottom of its gullet?"

"It lives on other fish, weaker than itself. It is the most formidable enemy of the small fry, which it devours in enormous numbers. But if it is the terror of the weak, it becomes in its turn the prey of a host of equally greedy eaters. At certain seasons of the year the cod gather in countless numbers and make long journeys to lay their eggs in favorable places. The famished denizens of the deep surround these schools of fish; the hungry inhabitants of the air soar over their course; the voracious occupants of the land await them on the shore. Man hastens to the spot to secure his share of the ocean manna. He equips fleets and sails in quest of the fish with naval armies in which all nations are represented; and what he catches he dries, salts, smokes, puts into casks, and packs in bales. Every year millions and millions of cod perish in this way, by man's fishhook, by the beak of birds of prey, and by the ferocious jaws of rapacious fish. With such extermination constantly going on it would seem that the end of the cod must be imminent; and yet there is no sign of it: the next year these fish resume their journey in as large numbers as ever."

"Nevertheless," said Claire, "their ranks must in the end become thinner by millions and millions."

"There is no sign of it, as I said before. A cod lays nine million eggs at a time! Where are the eaters that could put an end to such a family?"

"Nine million eggs!" exclaimed Jules; "what a family!"

"Just to count these eggs one by one would take nearly a year of eight or ten working hours daily."

"Whoever counted them must have had lots of patience," observed Claire.

"They are not counted; they are weighed, which is soon done; and from the weight it is easy to estimate the number when it is known how many eggs it takes to make a gram."

"Ha! how easy it is when you know how!" cried Claire. "What would have taken a year of tiresome work becomes the affair of a minute or two."

"One of the favorite rendezvous of the schools of cod is the neighborhood of Newfoundland, a large island of the seas that wash the eastern coasts of North America. Near this island is a vast extent of shallow water called the Newfoundland Banks. Thither in summer, attracted by abundant food, come myriads of cod from the depths of the northern seas. Thither, also, come fishermen of all nationalities.

"This is not the small fishing that you sometimes see on the banks of a river; the fishermen do not wait hour after hour under the shade of a willow for an ill-favored little carp to come and nibble at the hook baited with a worm, and count themselves lucky if they go home with half a dozen diminutive fish strung on a twig or lying in the bottom of a basket. Fishing in Newfoundland is a different matter: cod are caught by the ship-load. France alone sends out every year four or five hundred vessels, with crews aggregating fifteen thousand men, to the various fisheries conducted on a large scale; and among these fisheries that of the cod is chief and employs the most men. At the same time think of the Dutch, Danes, Swedes, English, Americans, and many others, all bound for the same fishing-grounds, their fleets manned by an army of fishermen, and you will gain some idea of the activity prevailing off the coast of Newfoundland.

"At daybreak the boats leave the ship and take their places, one here, another there, at the most promising spots. From both sides of each boat hang lines — stout cords of hemp carrying at the end an iron fish-hook baited with a small fish or with a shred from the entrails of cod taken the day before. The voracious codfish rush up at the sight of these dainties and greedily swallow at a gulp both hook and bait. The fisherman pulls in his line and the victim

138

follows, its gullet pierced by the fish-hook. Scarcely is the line baited again and thrown back into the water when another cod is caught. On both sides of the boat every man watches his lines and keeps on renewing the bait, throwing the line into the sea and pulling it in again with a cod at the end. By evening the boat is filled to the gunwales with fine large fish, still wriggling."

"Certainly that is a kind of fishing," said Claire, "that leaves no time for napping, with a line dangling idly at the end of one's rod; and when you do get a bite, it's no miserable little gudgeon, either, that has swallowed your hook."

"No, it is no miserable little gudgeon. The average length of a cod is one meter, and its average weight between seven and eight kilograms. Occasionally cod are taken that weigh as much as twenty or thirty kilograms. A fish of that size caught in one of our rivers would be the talk of all the country around."

"You say," Jules interposed, "that the hooks are baited sometimes with small fish, sometimes even with pieces of the entrails from the cod taken the day before. If they pounce like that on the remains of their own kind, codfish must be very greedy indeed! Other animals don't devour those of their own species."

"Their voracity is unequaled. With these fish the large gobble up the small, the strong devour the weak, without the least scruple and under no compulsion of extreme hunger, though they must often be hungry enough, as they are endowed with digestive powers that are truly astonishing. Moreover, if some indigestible prey, too bulky for comfort and swallowed too greedily, incommodes them, the cod have a quick way of getting rid of it: they reject the excess of food by vomiting."

"Oh, the horrid creatures! "cried Marie. "Their white flesh in its pretty layers doesn't correspond at all with their way of living."

"I do not deny it, but it is that particular way of living that has given us their savory white flesh as a highly esteemed article of food. And then this greediness that excites your disgust is not without its importance in the scheme of things. Think of a cod's family, of the nine million eggs laid by a single fish. If all those eggs hatched and the young were allowed to reach maturity, in a few generations the millions would become billions, and these latter would in turn multiply and become other billions, so that before long there would be no room in all the seas taken together for the codfish alone. Therefore these fishes must eat one another now and then, if only to offset this alarming multiplication. Man, birds of prey, large and voracious fish — all lend a hand in this work of extermination. And thus with immense slaughter the prolific cod is held down to reasonable limits within its ocean home instead of becoming a portentous multitude.

"Filled to overflowing, the boats return in the evening to their respective ships, where the preparation of the fish takes place. With a large knife one fisherman cuts off the heads; another slits the decapitated cods along the line of the stomach; a third takes out the entrails, being very careful to set aside

139

the liver; a fourth flattens the fish thus treated; and a fifth rubs them well with salt and piles them up."

"What do they do with the livers that were set aside! "asked Claire.

"They fill a cask with the livers and leave it exposed to the air. Soon decomposition sets in, the whole mass putrefies, and there rises to the top a greasy liquid that is known as cod-liver oil. This oil is carefully collected, for it is held in high repute as a medicine."

"I have heard of it, "said Marie. "They say it is detestable to take on account of its horrid smell of decayed fish. The way it is obtained accounts for its nastiness. Decayed fish-livers couldn't possibly furnish anything pleasing to taste or smell. But after all a person conquers his repugnance if the detestable remedy is really a cure."

Chapter Fifty-Four - Air

"Pass your hand rapidly before your face. Do you not feel a breeze! Now, instead of your hand use a good-sized piece of cardboard. The breeze becomes stronger. Try to run while holding an open umbrella behind you. You run with much difficulty and seem to be dragging not a light umbrella but a heavy load which opposes all its resistance to your progress, and soon your strength is exhausted.

"Whence comes that breeze, and what causes that resistance? It is the air that is answerable for both, the air in which we are all submerged like fishes in water. Is it not true that the hand, moved rapidly back and forth in the water, produces currents and eddies, with little weaves that ruffle the surface and beat against the banks! Precisely the same thing happens when air takes the place of water. Agitated by hand or cardboard, it is displaced, set in motion, and made to beat in successive waves against everything it encounters. Hence the puff of air that cools the cheek when a fan is used.

"If you undertook to run in the water and at the same time to drag after you a towel or even a handkerchief arranged so as to form a wide-mouthed pocket secured by the four corners, do you not think you would experience a resistance difficult if not impossible to overcome? In similar manner the fabric of the umbrella, arrested by the air, will not let you run fast. The more air you move, the more force you must exert. It is all simple enough. A wide piece of cardboard fans us much better than the hand alone; a large umbrella impedes our course far more than a small one.

"Have you ever noticed the reeds that grow in a running brook! They are kept in constant agitation. The dragon-flies or darning-needles, with great gauze wings and long green or blue bodies, that alight for a moment's rest on the tips of these reeds, have difficulty in maintaining their balance on that unsteady perch. Why are those reeds in continual motion!"

Marie hastened to reply: "The running water is striking against them all the time."

"Yes, that is plain. And the great trees, especially the tall poplars, that sway and bend in deep bows and then straighten themselves up, only to repeat the performance — what is it that moves them in that fashion? A giant's hand striving to uproot them would produce no such swaying to and fro. Obviously it is the air in motion that causes the trees to move, just as the water in motion makes the reeds move. Wind is air in motion, and its force is sufficient to snap the poplar tree, rend branches from the oak, and even overthrow solid walls.

"Invisible though it is, therefore, air is a very real substance; it is tangible matter no less than the water of the brook, the flood of the mighty river, the billows of the ocean. So long as it remains at rest we are unconscious of it; but let it be set in motion in great waves, we feel it as wind and are very sensible of its buffets. Without waiting for the next gale to convince us that air is matter, we can, with a little contrivance, examine at close quarters this substance that the hand cannot grasp or the eye see.

"Let us take a drinking-glass and plunge it into the water. It fills of itself. Now that it is full, let us hold it in any position we wish, but without taking it from the water. Whether its mouth be upward or downward or sidewise, the glass will remain full. As long as it is in the water we cannot empty it, not even by turning it upside down. And that is just what might have been expected; for what disposition could we make of its contents when there is water on every side to take the place of any that might flow out?"

"So much then being understood, what is the condition of a drinking-glass as we see it standing on the table among other preparations for dinner? Is it really empty as we say it is before filling it with water or wine? Does it contain nothing, absolutely nothing? If you insist that it is empty, I shall proceed to show you that it is full, full to the very brim. But full of what? Full of air, nothing but air.

"Being immersed in air, the glass has filled itself with air without any help from us, just as it would fill itself with water if it were plunged to the bottom of a well. Moreover, it remains full in any position, even upside down; for if any of the contained air should escape, the surrounding air would immediately take its place. Everything is ordered here just as we have seen in the case of a glass immersed in water.

"Accordingly, when we say of a glass or of a carafe, of a cask or a barrel, of a jug or a pitcher or any vessel whatever, that it is empty, the current expression used by us is not in accord with the exact truth. The vessel designated as empty is in reality full of air and remains full in whatever position it is placed.

"Returning now to the drinking-glass, let us plunge it into the water, holding it in a vertical position but with the opening downward. In vain do we push it deeper and deeper into the liquid as far as the arm can reach; this time the glass does not fill with water. Being already filled with air, as has

141

just been explained, it cannot be filled with anything else until that has been emptied out. The air imprisoned in the glass with no way of escape acts as an obstacle to the entrance of the water. Under this aspect, then, we see once more that air is real matter, capable of resistance and not yielding its place so long as there is no way open for it to go elsewhere.

"Let us release the prisoner. To do this we gently tip the glass sidewise while still holding it immersed. A diaphanous globule shoots up through the water and bursts on reaching the surface. Other globules follow, and still others, as we tip the glass more and more. They look like crystal pearls of incomparable clearness. These transparent globules, these pearls that make the water seem to boil, are nothing but air escaping from the glass in little spurts or bubbles; and thus the air is rendered visible despite its ordinary invisibility. We can distinguish it clearly from the water in the midst of which it makes its ascent. We follow its exit from the glass and note its upward passage in the form of bubbles; but, once arrived at the surface and mixed with the outer air, it escapes the keenest eyesight.

"We have just taken considerable pains to prove the existence of a substance unseen by anybody. Is it, then — this substance that we call air — something of importance? Assuredly it is: air is of the very first importance, since without it neither animal nor plant could exist. To give you an idea of the immense part it plays would require too much science and too much time. Let us confine ourselves to a more cursory treatment of the subject.

"When the fire burns low on the hearth and the glowing sticks of wood are turning dull, emitting smoke without flame and threatening to cease burning altogether, what do we do to revive the fire? We take the bellows and supply a blast of air. With each puff the dull coals turn brighter, the fire regains its vigor, and flames begin again to flicker. If generously fed with air from the bellows, the fireplace once more resumes its radiance.

"If, on the other hand, we wish to prevent a too rapid consumption of fuel, we partly cover the firebrands with a shovelful of ashes. Under this cover, which tends to keep out the air, the fire dies down somewhat. Indeed, it would go out entirely if the layer of ashes were to cover it completely and thus wholly exclude the air.

"Where on a cold winter day we gather around the glowing stove to warm our benumbed hands, we hear a subdued murmuring sound that tells us the fire is burning. We say then that the stove snores. This sound is caused by the inrush of air through the door of the ash-pit and its assault upon the mass of glowing firebrands, the heat of which it helps to maintain. The more air admitted, the hotter the stove becomes. If we wish to moderate the heat, we have only to close the door of the ash-pit. Thereupon, as air is then admitted only in small quantities through the joints and seams, the fire will slacken and the stove lose its red glow and turn black. The fire would die out entirely if no air whatever gained access to the mass of burning fuel.

"These examples make sufficiently clear to us that air is indispensable to all combustion; it revives a dying fire; it enters into the consumption of the

142

fuel, producing in the process both heat and light. Without air, no fire on our hearths, for wood and coal and other fuel bum only with the help of air. Without air, no light at night in our homes, for the illuminating flame of the lamp, of the candle, and of various other contrivances for dispelling darkness, goes out as soon as the supply of air fails."

Chapter Fifty-Five - Air (continued)

OF all the blessings we enjoy, the first place should be given to health, which is constantly menaced by diverse perils, and these are all the more to be feared when they are unknown. Let us learn what these dangers are and the means of avoiding them; then we shall be able to preserve our health, so far as its preservation depends upon ourselves.

"Prominent among the needs to which we are subject stand the need of food, the need of drink, and the need of sleep. But there is still another before which hunger and thirst, however violent they may be, lose their importance; a need continually born anew and never satisfied; a need that knows no respite and makes itself felt whether we wake or sleep; by day, by night, and all the time. It is the need of air.

"So necessary is air to the maintenance of life that it has not been left to us to control our use of it as we do in the case of food and drink. Unconsciously, and with no volition on our part, we admit the air to our lungs and allow it to play its wonderful part in our system. On air, more than on anything else, we live, our daily bread taking only second place. The need of food is felt only at comparatively long intervals; the need of air is felt uninterruptedly, ever imperious, ever inexorable. "To convince yourself of this, try for a moment to suspend the admission of air to your lungs by closing the doors against it, the nose and the mouth. In a few seconds you will be forced to end the experiment; you will begin to stifle and will feel that death would surely follow if you persisted in your experiment.

"All animals, from the smallest to the largest, are in like case with ourselves: first and foremost they live on air. Not even do those that live in water — the fishes and other forms of aquatic life — make any exception to the rule: they cannot live except in water containing a certain amount of air.

"There is a striking experiment in physics to illustrate this point. Some small living creature — a bird, for example — is put under a bell-glass from which the air is being gradually exhausted by means of an air-pump. As the supply of air diminishes under the action of the pump, the bird begins to totter, struggles in an anguish painful to see, and finally falls in the death agony. Unless air is quickly admitted once more to the bell-glass, the poor victim will be dead and nothing can restore it to life. But if air is admitted in time, it re-animates the bird. Again, if a lighted taper instead of a live bird be placed under the bell-glass, the flame is extinguished as soon as the air is withdrawn. The bird must have air if it is to live; the taper if it is to burn.

"What I am now going to tell you will explain briefly the reason for this necessity for air. Man and animals have a temperature suitable to them, a degree of warmth resulting, not from any outer circumstances, but from the vital processes within.

Clothing helps to retain this warmth, helps to prevent its dissipation, but does not supply it. Moreover, this natural warmth is the same under a burning sun and amid the frosts of winter, in the hottest of climates and in the coldest. Finally, it cannot be lessened without placing us in very serious danger. In the case of man its measurement on the centigrade thermometer is thirty-eight degrees.

"How is it that this warmth of the body is kept always and everywhere the same; and whence can it come if not from combustion? As a matter of fact there is going on in us a continual combustion, supplied with fuel in the form of food by our eating, and furnished with the necessary oxygen from the air we breathe. To live is to be burnt up in the strictest sense of the word; and to breathe is to burn. From time immemorial there has been in use, in a figurative sense, the expression, 'the torch of life.' We now perceive that this figurative expression is in reality the literal expression of the truth. Air makes the torch burn, and it also makes the animal burn. It causes the torch to give out heat and light, and it causes the animals to produce heat and motion. Without air the torch becomes extinct; without air the animal dies. In this respect the animal may be likened to a highly perfected machine set in motion by the heat from a furnace. The animal eats and breathes in order to generate heat and motion, receives its fuel in the form of food, and burns it up in its body with the aid of air supplied by breathing.

"We say that animals eat and breathe to generate heat and motion. That is why the need of food is greater in winter than in summer. The body cools off more rapidly in contact with the cold air outside, thus making it necessary to burn more fuel in order to keep up the natural warmth. A cold temperature, therefore, whets the appetite for food, while a warm one tends to dull its edge. The famishing stomachs of dwellers in the far North demand hearty food, such as fat meat and bacon; but the tribes of Sahara are satisfied with a daily ration of a few dates and a small portion of flour kneaded in the palm of the hand. Anything that lessens the loss of heat lessens also the need of food. Sleep, rest, warm clothing, all these serve to supplement the taking of nourishment and to conserve the natural heat of the body, even in a certain sense taking the place of nourishment. This truth finds expression in the common saying that he who sleeps dines.

"The fuel burnt up in us by the air we breathe is furnished by the very substance of our bodies, or more particularly by the blood, into which the food we digest is transformed. Of a person who applies himself to his work with excessive ardor we say that he burns the candle at both ends — another popular phrase that could not be bettered in its agreement with what is most assuredly known concerning the vital processes. Not a movement is made by

144

us, not a finger is lifted, without causing a consumption of fuel proportioned to the energy expended; and this fuel is furnished by the blood, which itself is maintained by the food we eat. Walking, running, working, putting forth effort, engaging in activity of any sort — all these do in a very real sense burn the bodily fuel just as a locomotive burns its coal in hauling after it the heavy burden of its long train of cars. Thus it is that exercise and hard work increase the need of food, whereas rest and idleness diminish this need.

"The coal in a furnace takes fire, becomes red-hot, and burns up, at the same time giving out heat. Soon there is nothing left of it but a quantity of ashes weighing much less than the coal consumed. What has become of the part represented by this difference in weight? I have already told you in my talk on combustion. It is no longer in the furnace in black lumps visible to the eye, but it is in the air in a form that the eye cannot see.

"Air, as chemistry tells us, is a mixture of two gases having very different properties the one from the other. These gases are oxygen, an active gas lending itself readily to combustion, and nitrogen, an inert gas with no tendency to combustion. In one hundred liters of this atmospheric mixture there are twenty-one liters of oxygen to seventy-nine of nitrogen. Now, in burning, the carbon of coal unites with the oxygen of the air, the two forming a gaseous compound called carbonic acid gas, which becomes diffused in the atmosphere. The part of the coal that is not carbon remains in the furnace, being insoluble in the atmosphere, and constitutes the ashes. All the carbon, then, disappears, seeming to undergo annihilation because we no longer see it, just as we cease to see the lump of sugar dissolved in water. This dissolution in oxygen, with the generation of heat, is called combustion.

"The incessant combustion going on in our bodies at the expense of the materials furnished by the blood is by no means comparable for violence with the combustion taking place in a furnace. It is a slow burning, somewhat like the spontaneous ignition of a damp hay-mow before it bursts into flame. It produces heat, but not enough to endanger the body as it would be endangered by undue proximity to a glowing furnace.

"In passing through a furnace and maintaining the fire therein, air changes its nature: its oxygen unites with carbon to form carbonic acid gas, which escapes through the chimney, while pure air is continually entering to take its place. Exactly the same process goes on in the combustion that keeps us alive. The lungs act as a pair of bellows, alternately filling themselves with air and emptying themselves. These alternating movements are known as inspiration and expiration. In the first, pure air is drawn in to burn up certain constituents of the blood and generate heat; in the second the air, after performing its office, is expelled, not the same in substance as when it entered, but impregnated with carbon and unfit for breathing, like the air escaping through the chimney from a furnace. The nitrogen in the air undergoes no change, but carbonic acid gas takes the place of most of the oxygen. In short, the breath from our lungs is essentially the same as the breath from a furnace."

Chapter Fifty-Six - Impure Air

AIR, on which our very existence from moment to moment depends, exposes us to serious dangers when it is vitiated with foreign emanations, with impurities, which, though perhaps harmless when inhaled in only a breath or two, are fraught with peril if their admission to the lungs continues. Breathing is never suspended, day or night, and anything that disturbs it even but slightly causes uneasiness at first and then, before long, grave danger. We are careful to have our food clean; we ought to be still more careful to have our air pure, its part in the maintenance of life being more important than even that of food. All air injurious to health from any cause whatsoever is called impure air.

"This impurity may be brought about in various ways, especially by the mixture of air with other gases, some of them dangerous merely in that they cannot take the place of oxygen in the combustion unceasingly going on in our bodies, others bringing peril with them in the form of poisons that infect the blood. Foremost among these latter is the carbonic oxide, or carbon monoxide, as it is also called, which all our devices for heating generate in greater or less quantity, and which is therefore a serious source of peril. This terrible gas is produced in our very homes.

"Carbon can burn with two different degrees of completeness; that is, it can combine with a single or double portion of oxygen. Completely consumed, it gives carbonic acid gas, the gas produced by our breathing; half consumed it gives carbonic oxide. Let us turn our attention to the flame of a lighted candle. Just at the base of the flame we see a narrow band of beautiful blue. On the top of slowly burning coal may be seen little tongues of flame having the same blue color. That is the distinctive sign of the gas we are considering; those blue flames are produced by the complete combustion of carbon in the formation of carbonic acid gas. But before this final process the gas generated by the semi-combustion of carbon is as invisible and subtle as the air itself.

"Carbonic oxide is odorless, so that we remain unsuspicious of its presence, to our great peril. Inhaled even in a very small quantity, it poisons the blood, causing at first a violent headache with general discomfort, then vertigo, nausea, extreme weakness, and finally there may be a loss of consciousness. However short the duration of this state, life is imperilled. In my talk on combustion I warned you against this terrible gas, pointing out what precautions should be taken in respect to heating apparatus, the braziers used in laundry-work, and even our modest foot-stoves. All these, unless proper provision is made for the circulation of fresh air, expose us to serious danger. There is no need to dwell further on this subject; I have already said enough.

"Carbonic acid gas, which represents the final stage in the combustion of carbon, is not a poison like carbonic oxide; our lungs always contain some of it, since it is in every breath we exhale; but, though it is not a poison, it is un-breathable, being by no means a substitute for the oxygen it has replaced in the composition of the air. Now, carbonic acid gas is produced in abundance all around us; in the first place, during every moment of our lives, by our own breathing; secondly, by the combustion that goes on in our heating applianc-es; thirdly, and less frequently, by fermentation.

"I shall not here take up again the subject of our ordinary means of heating and the dangers to our health that lurk therein and call for vigilant precau-tions on our part. It is a topic already sufficiently familiar to us. Let us say a few words about movable stoves.

"The ordinary stove [1] has its undeniable merits. It is the best heating ap-paratus in respect to facility of installation, economy of fuel, and conserva-tion of heat. But it is defective from the hygienic standpoint, especially in a small room occupied by many persons; it does not provide adequately for the renewal of the air, which it is capable of poisoning with its carbonic oxide.

"A form of stove that should be absolutely forbidden in our dwellings is the so-called American or movable stove, which is furnished all ready to be set up and is installed perhaps first in one room and then in another, without proper precautions as to its adaptability to the apartment to be heated. By its mode of slow combustion it generates carbonic oxide in great quantities, and this may escape into the room at night and bring death to the sleepers. Too many accidents have already demonstrated the peril lurking in this form of heater.

"Fermentation, or the decomposition that goes on in the sweet juice of grapes in the process of turning to wine, produces carbonic acid gas in abun-dance. Hence it would be foolhardy unless assured beforehand of finding ad-equate ventilation, to make one's way into the vault or cellar where must is fermenting; and it would be still more foolhardy to descend into a wine-vat even after the wine itself has been drawn out. Carbonic acid gas may be there, forming a layer or stratum of some depth, since it naturally sinks to the bottom because of its being heavier than air. Hardly has the rash adven-turer entered this layer, entirely undetected by the eye, when he falls uncon-scious, as if struck by lightning. Succor in this conjuncture is fraught with danger, and when it arrives it is often too late.

"The lesson is plain: the most elementary prudence calls for. circumspec-tion in entering any room or other enclosure where carbonic acid gas from fermenting wine may be present. Before entering there should always be a preliminary testing of the atmosphere. A lighted paper attached to the end of a long pole should be introduced into the wine-vat and into the remotest corners of the cellar. If the paper burns as usual, the atmosphere is free from danger; if it burns dimly, smokes, or, surest sign of all, goes out altogether, carbonic acid gas is certainly present. Until ventilation has removed the dan-ger let no one venture where the paper refuses to burn.

"Similar precautions should be taken in the case of recently opened crypts, deep excavations in abandoned quarries, and unused wells. Their atmosphere, often vitiated by carbonic acid gas, should first be tested with a lighted paper.

"After entering the lungs air gives up a part of its oxygen to the blood and receives in exchange an equal volume of carbonic acid gas produced by the combustion that takes place in the body. Hence the breath exhaled from the lungs is less vivifying than normal air. No argument is needed to prove that respiration cannot continue indefinitely in an atmosphere not subject to renewal. The ordinary proportion, twenty-one per cent, of oxygen in the air is never completely exhausted; but any considerable diminution, with the accompanying substitution of carbonic acid gas, is enough to render breathing difficult and, before long, dangerous. Placed under a bell-glass with no provision for renewing the air, any living creature will succumb in time, its duration depending on the rate of respiration.

"One word more: in its passage through the lungs the breath becomes laden with noxious emanations, the invisible refuse of the human organism in its unceasing process of destruction and reconstruction. Man's breath is pernicious to man. Air is vitiated by merely remaining in the lungs, where it loses a part of its vivifying element.

"For these various reasons it is important that strict attention be paid to the renewal of the air in dwellings, especially in the rooms used for sleeping, these latter being usually small and almost always kept carefully closed for the sake of warmth and quiet. Alcoves and bed-curtains protect us, so that we are shut in with a limited supply of air which all night long undergoes no renewal, whereas in the other rooms of the house there is during the day a constant circulation of air through the frequent opening of windows and doors. When we awake in the morning the air about us cannot but be impure. Let us then open our bedroom windows as soon as possible and allow the pure outside air to flood the chamber and replace the unwholesome atmosphere formed during our sleep. Let us also admit the sunlight, another powerful vivifying agent, that it may penetrate with sanitary effect the depths of our alcoves and the darkness created by our drawn curtains.

"Under ordinary conditions air circulates about us in such abundance that we hardly have to pay any attention to its quantity; but this is not so in a closed room, a dormitory for instance, where a number of persons pass the night. Then it becomes necessary to provide a sufficiency of air for each one breathing within this limited space. Now, there passes through the lungs of every one of us about ten thousand liters of air per day, or four hundred and fifty per hour, which amounts to nearly four cubic meters for a night of eight hours. But since air is vitiated by respiration long before its oxygen is notably lessened in quantity, it is advisable to multiply this four by ten at least and to provide forty cubic meters of space for each person occupying a sleeping-chamber in which the air, all night long, is not renewed. Indeed, if double this allowance is made, the demands of hygiene would be no more than satisfied.

"Eggs that are no longer fresh, and in which decomposition has begun, give out a noisome odor known well enough as the smell of rotten eggs. They tarnish any silverware that comes in contact with food containing them. These two effects proceed from the same cause, sulphureted hydrogen, a compound of sulphur and hydrogen. This gas is not only nauseating; it is also, a more serious matter, highly poisonous, being comparable with carbonic oxide in its harmful qualities. Whoever inhales it quickly succumbs.

"Now it so happens that sulphureted hydrogen is a constant product of the decomposition taking place in excrementitious matter, so that care is necessary that dwellings and especially sleeping-rooms be at a safe distance from all such sources of infection, and that any possible danger therefrom be counteracted by abundant ventilation.

"Nor is sulphureted hydrogen the only injurious product of decomposition. Everything that decays gives forth exhalations, whether malodorous or not, that cannot but injure our health. Therefore let us put a safe distance between our dwelling-houses and our stables, hen-coops, rabbit-hutches, dung-heaps, and other similar menaces to human welfare.

"In the country there is shown a somewhat excessive scorn of these dangers, but this carelessness is usually counterbalanced by plenty of ventilation. Doors and windows imperfectly closed, badly fitting, gaping with cracks, and often wide open, allow free access of fresh air to every nook and comer. Nevertheless every deposit of refuse too near a well threatens the health of the whole household using that well."

[1] The European built-in stove of tile or brick is here meant. — Translator.

Chapter Fifty-Seven - Germs

"WHEN a sunbeam penetrates the twilight of a darkened room its course is defined by a straight shaft in which innumerable corpuscles, rendered visible by the bright illumination, are seen whirling and eddying in constant though gentle motion. Outside of this shaft the air, though it appears to us perfectly limpid, is laden in the same degree with similar particles of dust. On account of their smallness these atmospheric impurities escape detection; but let a ray of sunshine illuminate them and turn each one into a point of light, and straightway they become visible, at any rate the largest ones. Others, and these form the greater number, defy the scrutiny of the sharpest eye even in a ray of light.

"Now, what are these atoms, visible and invisible? They are made up of an inextricable mixture of a little of everything. There are mineral particles raised from the ground by the wind, coal-dust from the smoke emitted by our stoves and fireplaces, tiny shreds of wool and cotton worn away by the friction of our clothes, minute fragments of wood from the wainscoting, and loosened particles of paint. In short, the pulverized product of all the wear and tear that goes on around us is here represented.

"So far there is nothing remarkable and certainly nothing dangerous unless in exceptional circumstances, when this atmospheric dust comes from poisonous substances. Our interest in the subject increases when we learn that there also float in the air, in prodigious multitudes, all sorts of germs of animalcules, seeds of the lowest forms of vegetable life, in most instances too small to be seen without the aid of a good microscope.

"Let us consider the simplest case, that of mold. Every one knows that a slice of any fruit, a melon for example, if left exposed to the air, no sooner decays than it is covered with a long silky down. This down, composed of ramified filaments standing up against one another, is a vegetable growth belonging to the same class as the mushroom. Its common name is mold. Whence comes this plant, so curious if examined closely, with its little black heads full of spores? Is it engendered by decay?

"That is the general belief; but let us not be deceived. Decay does not engender anything; it is not a cause but a result. The slice of melon in decaying does not create the mold; on the contrary, it is the mold that induces the decay of the slice of melon, at the expense of which it develops. This lowest form of vegetable life has its origin in a germ, we may call it a seed, just as an oak has its origin in an acorn. Every living thing, animal or plant, without any exception, is derived from a previous living thing of like sort, which has furnished the germ or seed for the new life. Life is always the product of life, never of decay.

"Accordingly, the mold must have been sown. But by whom or what! Evidently by the air, for air is the only thing that has come in contact with the slice of melon. The conclusion is obvious: there are in the air, floating unseen amid the multitude of other microscopic particles, the germs of mold that induce the rotting of fruit; they are there in immense numbers, for the growth of mold is very thick; they are everywhere, for in whatever part of the house the slice of melon is left, it is attacked by mold; and, finally, there are many different kinds, each one attacking the vegetable or animal substance it likes best.

"These germs come from former molds whose invisible and innumerable seeds were thrown into the air by the bursting open of the ripened fruit of the mold. Almost without weight, and wafted this way and that by the slightest current of air, they sooner or later fall upon some body favorable to their germination, a slice of melon or something else. The sowing is accomplished and the plant grows.

"Held in the air and borne hither and thither by it are similar multitudes of germs representing organisms known as animalcules, infinite in variety, and all too minute to be seen without the aid of a microscope. These animalcules are called infusoria because the simplest way to obtain them is to infuse in water any substance, animal or vegetable. Let us put to soak in a little water a few pinches of hay chopped up fine, or some bits of grass, no matter which. In a few days, especially in the heat of summer, the most curious population will be found swarming in the liquid. A drop of this water no bigger than a pin's head shows us under the powerful eye of the microscope a startling spectacle. In the ocean of this drop confined between two thin plates of glass, there come and go, swim and plunge and rise again by the aid of their cilia, infusoria of many varieties, all in unceasing motion. Some are flattened and oval in shape, somewhat resembling a certain sea-fish, the sole; others are bristling globules, whirling rapidly; still others, attached to some bit of foliage by a spiral thread, present from above the shape of a bell, the mouth of which, fringed with cilia in rapid vibration, gives the appearance of a wheel revolving swiftly. Then suddenly the spiral thread tightens, the opening of the bell closes, the cilia cease to vibrate. A tiny prey has just been caught and the infusorium contracts to digest its victim at leisure.

"Let us pause here in our description of these animalcules and turn our attention to the origin of the infusoria.

"This aquatic population having its home in water in which grass or other organic matter has decayed can come only from germs brought by the air. Let us boil a little hay in water and pour the still hot and limpid decoction into a glass. The liquid, when freed from the last remnants of hay, is perfectly clear, with a slightly yellowish tinge. But, behold, in a few days it becomes clouded, turbid. Examined closely under a microscope, it is seen to be peopled with infusoria. The germs of these animalcules were not furnished by the water and hay used, for if they contained any, which might well have been the case, the act of boiling would have killed them beyond resuscitation. Let us bear this fact well in mind, as we shall soon find important applications of it; no live thing, even if it be in the form of a germ, can withstand the temperature of boiling water. Plant, animal, eggy germ, seed, all perish in the heat required for raising water to the boiling point. Our boiling-hot decoction, therefore, did not contain anything having life. If, then, a few days after cooling off, it is found to be teeming with life, these organisms can owe their origin only to the dust of the air, rich in infusorial germs.

"Should any doubts remain on this point, the following experiment will dispel them. The infusion is poured into a glass flask, the neck of which is then melted and drawn out into a fine tube. The liquid is now raised to the boiling point in the body of the flask. Steam rises and as it escapes in a jet through the small opening of the extended neck it drives out all the air, after which the flask is hermetically sealed by melting the tip of the neck. Henceforth no infusoria can by any possibility make their appearance. For years and years the decoction of hay will remain perfectly limpid, developing not the slightest cloudiness, and microscopic examination will prove that the clear liquid contains nothing capable of producing life without intervention from outside. But let the tip of the neck be broken and air enter, and very soon the usual infusoria will appear.

"Below the infusoria are microbes, smaller in size, of much simpler structure, and apparently belonging to the vegetable kingdom. Microbes are the infinitely small in living form. A thousand of them placed end to end would in most instances measure scarcely a millimeter. There are some that are visible only under the most powerful microscope. Thus examined, they appear as bright points, constantly trembling and of various shapes, some oval or rounded, others rod-shaped, and still others bent or curved like a comma. They are everywhere, in numbers that defy counting; they are in the air, the water, the ground, in decaying matter, in the bodies of animals, and in our own bodies. They lay claim to everything.

"What part do these infinitesimal organisms play in the order of things? They play a very important one. Let us give two examples.

"An animal dies. Its body decays and is soon resolved into its primitive elements, which are seized upon for the nourishment of growing vegetation. The putrefying flesh is converted into flower, fruit, grain, nutritive matter. What agency effects this wonderful transformation? It is the microbe, which clears away dead matter and restores its elements to the realm of the living. By developing and multiplying they induce decay, which gives back to life's workshop materials otherwise unavailable for use. Without their intervention the work of life would be impossible because the work of death would be incomplete.

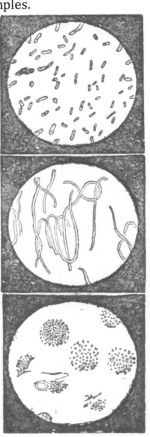

Microbes (much enlarged)

"As another instance take grape juice in the process of fermenting and becoming wine. The liquid heats as the result of its own internal activity; it begins to foam, and big bubbles of carbonic acid gas are formed, until at last there is developed that winy flavor which succeeds to the sugary taste of the earlier stages. This process is called fermentation. What induces it and thus gives us wine? It is a microbe, the same as that we find in yeast. In order to get nourishment for itself and to multiply until its numbers defy computation, this yeast microbe decomposes the sugar in the grape juice, resolving it into alcohol, which remains in the liquid, and carbonic acid gas, which escapes. Such is the secret of the making of wine, beer, and other fermented drinks.

"Among the various tasks performed by microbes let us henceforth remember putrefaction and fermentation.

"These two infinitely small destroyers, one of which makes alcohol out of sugar, and the other reduces a dead body to dust and gas, warn us that other microbes carrying on their work of demolition at the expense of our own organs may by their prodigious multitude give rise to dangerous diseases. One of them, in fact, produces cholera, that terrible epidemic the very name of which terrifies us; another causes typhoid fever, which every year claims thousands of victims; and still others, each according to its aptitude, engender various ills which science is every day making known to us in increasing numbers. The decay of our teeth, with the extreme pain caused by it, is due to a microbe; the slow wasting away of a consumptive's lungs is traced to a microbe; the big purple carbuncle that causes so much suffering, this too is the work of a microbe. Enough of these tortures for the present. You can see that of all our enemies the infinitely small is the most terrible.

"Microbes are everywhere, we say, especially where filthy conditions prevail, whether in the air or in the water. The air in a hospital ward contains more microbes than that outside; the air of cities, where we live piled one on another, so to speak, has far more than the air of the country; the air of low plains carries a greater number than that of the uplands. High mountain air has none at all. There in truth may be found the pure atmosphere so conducive to health.

"Water is even richer than air in microbes, especially when it is defiled with sewage. It is estimated that such water may contain various kinds of microbes and their germs to the frightful number of one hundred million per liter. Not all of these, it is safe to say, are harmful; far from it; but in so immense a multitude there must certainly be some bad ones."

Chapter Fifty-Eight - The Atmosphere

"As we have seen, all animal and vegetable life has need of air, and air is supplied in inexhaustible volume. It forms around the earth a continuous envelop known as the atmosphere and having a thickness of at least fifteen leagues.

"It is a veritable ocean of air, but an ocean without shores, an ocean holding in its lowest depths all that lives and moves upon this earth of ours. In its upper reaches it extends far beyond the loftiest mountain peaks and occupies regions of space that no bird in its most daring flight has ever visited; and it makes its bed upon the dry land of the continents and upon the waters that encircle them, these latter constituting another ocean far heavier and denser, the abode of the aquatic population of the globe.

"In the daytime we see above our heads a boundless blue vault which we call the sky; but this vault is one in appearance only, owing its seeming existence to the atmosphere. To account for the azure cupola of the sky, note that substances very slightly tinged with color do not show this color until they are seen in masses of great thickness. A pane of glass looks colorless, yet if viewed edgewise it is found to be of a delicate green hue. In the first position there is nothing in the thin pane to arrest the eye; in the second, the considerable body of glass presented reveals the green color.

"Similarly, the water in a bottle, being of small volume, appears colorless; but if we look at a great mass of water, such as a lake or the ocean, it is seen to be tinged with either green or blue. The same holds true of air: colorless and therefore invisible where a volume of only moderate thickness is concerned, it becomes visible and shows its delicate tint of blue if the thickness be considerably increased. Thus it is that the enormously thick layer surrounding the earth assumes the appearance of an azure vault.

"Since it is matter, air must have weight; and in fact it has — very slight weight, it is true, but more than that of many other substances. Suppose we had a hollow cube measuring one meter each way. The air contained in this cube would weigh one kilometer and three hundred grams. A like volume of water weighs one thousand kilograms, or seven hundred and sixty-nine times as much.

"It is in the atmosphere, now higher, now lower, that the clouds float; and it is in the atmosphere that smoke rises and is dissipated. Why do clouds remain at a considerable height, and why does smoke rise? Because they are lighter than air. A piece of wood, if forced down to the bottom of a body of water and then released, immediately comes up again of its own accord; it does so because it is lighter than water. Exactly similar is the behavior of clouds and smoke, which are lighter than air, in the atmospheric ocean in which they are immersed. If there were no atmosphere, smoke would not leave the ground and clouds would trail along the surface of the earth. If there were no atmosphere to offer resistance to the strokes of their wings, birds could not fly.

"To protect us from the cold we have clothing. The terrestrial globe likewise has its thick blanket under which is preserved for some time the heat received from the sun during the day; it has its atmosphere, a cloak of air fifteen leagues thick. Without this protection, which plays a part similar to that of our eider-down bed-coverings in retaining heat, the earth would undergo, every night, a cooling-off that no living creature could withstand.

"With any diminution in the thickness of this envelop of air there is a corresponding decrease in the protection it furnishes, just as is the case in respect to the clothes we wear. Hence it is found that the cold increases rapidly in the upper regions of the atmosphere, because the protecting covering is thinner in proportion to the depth of atmosphere below. Thus we understand why very high mountains are covered with snow the year around, not even excepting summer: their summits, less protected by the atmospheric blanket than the surrounding plains, are subjected to a more rigorous cooling-off at night.

"A considerable mass of cotton may be compressed in the hands until it becomes a small ball. In like manner air is compressible: it becomes denser and occupies less space in proportion to the pressure exerted upon it. With this truth in mind let us consider the atmospheric envelop in its entire thickness. The layer next to the ground bears the weight of all that is above; it sustains the greatest pressure and therefore, volume for volume, is the richest in matter, just as the most compact ball of cotton is the one containing the most of that material. Without entering into further details we see that the atmosphere becomes less and less dense as we ascend, because it has less and less of superincumbent atmosphere to support.

"Living at the bottom of this atmospheric ocean, we breathe its lower layers, the air of which, denser than elsewhere, meets the needs of our lungs. If we ascend three or four thousand meters we find the air thinner and respiration difficult and inadequate. Higher still, we experience something worse than discomfort; we face very serious danger. Finally, at a height not really tremendous in itself strength fails, the mind wanders, and sudden faintness supervenes, followed by death.

"Though life ceases for lack of sufficient air, the atmosphere is still there, its upper layers extending far beyond the elevation here referred to; nevertheless it has not the degree of density required for sustaining life. Only in the lower layers of the atmosphere, then, is the air suitable for breathing; at a greater height all life ceases. Of course no bird mounts to those desert spaces; nor, in fact, would its wings find there a sufficiently resistant atmosphere to admit of flying.

"Freezing cold, prostrating physical discomfort, and at last sudden death — these await the daring adventurer who attempts to scale the vault of heaven. Let us then stay below, in the lowest depths of the atmosphere, the only abode suited to our needs; and if we are seized with a desire to know more in detail what goes on up yonder, let us be content to hear the report of those audacious explorers who have penetrated the upper regions of the atmosphere.

"A balloon ascends to heights unattained by the loftiest mountain peaks. The enormous globe of woven fabric is inflated with a gas called hydrogen, like ordinary air in its invisibility, but far rarer and lighter, weighing only a hundred grams to the cubic meter, or one thirteenth as much as air. Thus rendered buoyant by this tenuous gas, the balloon ascends because its total weight is less than that of a like volume of atmospheric air.

"It carries heavy weights, very heavy weights, it is true, notably the aeronaut, the one who makes the aerial voyage; but it rises in spite of this. How is it to be explained! It is very simple. Recall once more the piece of wood forced to the bottom of a body of water and rising of its own accord as soon as released. A piece of lead would not have acted thus, because lead is heavier than water; but with outside help it will rise readily enough. Attach to it a piece of wood or, better, of cork, making sure that the piece is large enough, and the whole will return to the surface after being forced down into the water and then released. The wood or cork, being lighter than lead, will pull up the latter with it as it rises. Exactly so does the balloon conduct itself: the very light gas inflating it carries aloft, as it ascends, the heavy objects in the car or great wicker basket suspended from the gas-bag.

"When he wishes to come down again the aeronaut opens a valve by pulling a cord hanging within reach of his hand. A little hydrogen escapes and ordinary air takes its place, whereupon the balloon, rendered so much the heavier, begins to descend, slowly or rapidly according to the amount of gas discharged. With this explanation of the principles governing the balloon's ascent and descent let us turn to the narrative of an English scientist, James Glaisher, who in September of the year 1862 attained the greatest elevation yet reached by man.

"'We left the earth,' he says, 'at one o'clock in the afternoon, in a mild temperature. Ten minutes later we were floating in a dense cloud which shrouded us in impenetrable gloom. After passing out of this layer of fog the balloon rose to a region flooded with light, the brilliant sunshine giving to the sky an extraordinarily vivid tint of blue.

"'Above our heads we had nothing but the azure of the firmament; under our feet lay a widely extended surface of clouds arranged in imitation of hills, mountain chains, and isolated peaks, all of resplendent whiteness. One might have mistaken it for a mountain scene covered with snow of incomparable purity. As we ascended we caught momentary glimpses of the earth through occasional openings in the clouds.

"'In twenty -five minutes the balloon had carried us up four thousand, eight hundred meters, which is very nearly the height of Mont Blanc, the loftiest mountain peak of Europe. A like ascent on foot would have cost us several days of extremely wearisome climbing. The temperature had by this time fallen very low, and ice was forming on the balloon. Another upward spurt raised us to the height of eight thousand meters, where the severest of winter cold prevails; and still we continued to ascend.

"'When we had reached an altitude of eleven thousand meters, or almost a league beyond the highest mountain in the world, my assistant, Coxwell, noticed that the valve cord was entangled in the rigging of the balloon, and he climbed up to disengage it.

"'Just then my right arm became suddenly paralyzed. I tried to use my left, but it too was paralyzed. They both refused to obey my will. I then attempted to move my body, but with so little success that I seemed to have no longer any body at all. I essayed at least to read the marks on the instruments, but my head fell over, inert, on my shoulder.

"'I had my back against the edge of the car, and in this position I looked up at Coxwell, who was engaged in disentangling the valve cord. I tried to speak to him, but could not utter a sound. Finally thick darkness came over me, my eyesight having in its turn become paralyzed. Nevertheless I was still perfectly conscious. I thought I needed air and that I should die unless we could very soon manage to descend. With that I lost consciousness just as if I had suddenly fallen asleep.' [1]

"'A minute more of this swooning fit and it would have been all over with Glaisher. Coxwell climbed up into the cordage amid long icicles hanging from the balloon. He hardly had time to free the valve cord: the extreme cold had seized him and his hands, benumbed and blue, refused to do their office. He had to descend into the basket by clinging to the ropes with his elbows.

"Seeing Glaisher lying motionless on his back Coxwell at first thought his companion was resting, and he spoke to him, but received no reply. The prostrate man's silence indicated that he had fainted. Coxwell then undertook to succor him, but paralysis and insensibility were fast overcoming the assistant also and he could not drag himself to the dying man's side, whereupon he at last became convinced that they must descend without delay if they were not both to perish in a very few minutes.

"Fortunately the valve cord hung within his reach; but being unable to grasp it with his hands, benumbed as they were with cold, he seized it between his teeth and after a few tugs succeeded in opening the valve. Immediately the balloon began to descend. Before long, in an atmosphere less cold and less rarefied, Glaisher recovered consciousness and gave his attention to the frozen hands of his companion.

"They regained the earth safe and sound, both men, but with no desire for further ascents of a like perilous nature. Far more serious was the issue in the case of three French aeronauts who, some years later, wishing to add to our knowledge of atmospheric conditions, made an equally daring ascent. When their balloon came down again two of the rash explorers were dead, stiffened by the cold and suffocated by the insufficiency of the upper air, while the third, saved as by a miracle, was at his last gasp. Knowledge is sometimes bought very dearly. Science has its heroes and martyrs."

[1] The paragraphs ostensibly quoted from Glaisher's narrative are in reality a very free paraphrase, much abridged. For the author's full account see his "Travels in the Air," Part I. Chapter 3. — *Translator.*

Chapter Fifty-Nine - Evaporation

MOTHER AMBROISINE had just been washing some handkerchiefs. After soaping them she rinsed them in clear water, and then wrung them out as dry as she could by energetic twisting and squeezing. That done, were they dry enough, and could they be used just as they were!

This question Uncle Paul put to the children, and they all agreed that the handkerchiefs were still very wet, that in fact they held an amount of water exceeding their own weight. What, then, was to be done in order to make them as dry as linen must be before it can be used!

"You all know," resumed Uncle Paul, "what Mother Ambroisine will do to the handkerchiefs: she will hang them on a line in the sun and -in a light current of air, if possible. If conditions are favorable, if the sun is warm and a gentle breeze is blowing, the handkerchiefs will soon dry. Then they will only have to be ironed and put away in one of the bureau drawers.

"If the sun does not shine and it is cold, with no air stirring, the drying will take longer. But finally, sooner or later, the handkerchiefs will all get dry; they will lose the water they soaked up at the beginning.

"Let us take another example. We will set a plate of water in the sun. In summer, when it is warm and clear, the water will all disappear between

157

morning and evening, and the bottom of the plate will be found quite dry. In winter wait a few days, a few weeks perhaps, according to the weather, and the same result will be attained: the plate will, in the end, become quite empty although not a single drop has escaped by leaking.

"But is it necessary, after all, to have recourse to any such experiments as these? Will it not suffice to recall to mind what each one of us has witnessed over and over again? Who does not know the little puddles of water, the stagnant pools, that collect in ruts and hollows whenever there is a fall of rain?

"You go by when the pool is full. Ducks are dabbling in it and frogs croaking; black tadpoles, toads-that-are-to-be, sun themselves on the banks, their backs exposed to the noonday warmth, their bellies in the tepid mud. Strange plants, *confervae,* as they are called, display their long tufts of sticky green filaments.

"You go by again a little later and no more ducks are dabbling, no more frogs croaking, no more tadpoles frisking, no more *confervae* showing their verdure. All have vanished. The pool is dry. Doubtless the soil has drunk up, little by little, at least a part of the stagnant sheet of water where the toad "s black family was disporting itself and where the little ducks came waddling in single file to take their first lessons in swimming; but in many cases this slow infiltration of the water into the ground cannot account for the pool's disappearance.

"It may be that the bottom of the natural basin in which the rain-water has collected is formed of compact earth — of clay, for example, which is absolutely impervious to water — or it may be the bottom is of solid rock which, by its very nature, allows not the slightest infiltration.

"How, then, has the pool disappeared? What has become of the water it contained, if the earth has not drunk it up? There were, perhaps, thousands of liters, and now it is all gone. A thirsty chaffinch would not find enough water there to wet its throat. What, too, became of the plateful of water set in the sun, and the moisture in the linen washed by Mother Ambroisine?

"To find the answer to this question, which will lead us farther than you think, it is enough to recall what happens when a pot full of water is put on the fire. The liquid first gets warm and then begins to boil, while from the pot there burst turbulent jets of what looks like white smoke, hot and damp, and known to every one as steam.

"Now, this white smoke, this steam, is water, nothing but water; yet water under another form, water that, instead of running or dripping, expands in the air, floating there as light and thin as the air itself, and becomes dissipated until not a particle of it is visible.

"Watch a puff of steam coming out of the pot. You can see plainly enough the jet of white vapor at the mouth of the vessel, but a little higher up you see nothing whatever, the white puff having become dissipated in the air and being henceforth lost to sight. We no longer see the steam; nevertheless it still exists, whether it has been dispersed through the room, whence it will

158

escape by the doors and windows, or has been carried up the chimney by the current of air ascending from the fireplace.

"Thus the pot loses its contents through its mouth; little by little it empties itself from above; it yields its water to the atmosphere under the form of vapor. The hotter the fire the more rapid the loss. Always losing in this way and receiving nothing, the pot must sooner or later become dry. If the cook does not watch it and, before it is too late, replace the water that has boiled away, the vegetables she has put in it to cook will burn.

"What are we to conclude from this instance of the pot of water on the fire? This: the heat reduces water to vapor or, in other words, to something as thin and invisible as air itself. I insist on this word invisible because — note this well — the white smoke we distinctly see rising from the pot is not yet real vapor.

"Let us call it, if you like, imperfect vapor, or visible vapor, or mist. But when the white smoke has been dissipated in the air and become so thin and limpid that the eye can no longer detect it, then it is real vapor.

"What the intense heat of the fire accomplishes in a short time the sun also effects, but more slowly. It is the sun's heat, then, that dries up the pool by changing its water into vapor; it is the sun's heat that dries the plate by reducing its contents to vapor; it is the sun's heat that dries the linen hung on the line by converting the moisture it holds into vapor.

"Vapor from the pool, from the plate, from the linen, from the boiling pot, all goes into the air and floats there invisible, driven hither and thither by the least puff of wind. The greater the heat, the more rapid the transformation of water into vapor, and also the greater the air's capacity for receiving this charge of invisible moisture.

"That is why the duck-pond dries up sooner in summer than in winter, and why the linen that dries so quickly on a hot day is very slow in drying on a cold and dull one.

"But, whatever the temperature, air cannot receive an unlimited quantity of vapor. When it has a certain amount it becomes too damp to absorb a fresh supply of moisture.

"A perfectly dry sponge readily drinks up a certain quantity of water; already wet, it can take up only a smaller quantity; and if entirely saturated, it will take up none at all. A pile of dry sand, with its base resting in water, gradually becomes damp to the very top, and when it is thus wet through it cannot absorb any more water.

"Air behaves in the same manner: in a dry state it readily absorbs vapor; saturated to a certain point, it will receive no more. It is a soaked sponge in its powerlessness to drink up more water. So you can easily understand why air in motion, that is to say wind, accelerates the drying of linen and the disappearance of the water in a pond.

"The more humid the air, the less rapidly will it receive vapor, the formation of which is thus arrested; but if the air in contact with the pond, or with wet linen, or with any sheet of water, be constantly renewed by a breeze, the damp air is succeeded by dry, which in its turn becomes charged with vapor and gives place to other air, and this also carries on the drying process. Thus the transformation of water into vapor proceeds uninterruptedly.

"Let us sum up what we have just learned. Heat changes water into vapor, that is to say into something light and thin, which floats in the air and becomes dissipated until it is as invisible as the air itself. This change is called evaporation. Water evaporates at any temperature, but the more rapidly the greater the heat."

Chapter Sixty - Humidity in the Atmosphere

"AFTER a good rain the fields are muddy; thousands and thousands of threads of water running down the slopes have filled the ditches or collected in puddles and pools; the foliage of the trees shows a glossy green and shines as if the watery coating left by the storm were so much varnish; at the end of each leaf, at the tip of the tiniest blade of grass, trembles a drop of water, flashing like a diamond in the sunlight.

"But wait a few days and if the sun is hot and the wind blows a little, there will not be left a trace of this shower that gladdened the farmer's heart. The earth will have become dusty again, in the woods the little cushions of moss that but lately were so exquisitely fresh will have shriveled up and faded, the leaves of the trees will hang limp and lifeless, the puddles will be dry, and the mud will have turned to dust.

"What has become of the quantities of water discharged by the storm? The soil drank part of it, that is certain, to the benefit of the growing vegetation; but the air — insatiable toper! — also took its share, and a large one. The water that fell in the form of rain from the upper air has by evaporation returned to the atmosphere whence it came; the air has taken back what it gave to the soil for a little while. The drops that sparkled at the ends of the leaves have returned, invisible, to the immensities of the atmosphere without touching the earth. In short, evaporation has dissipated the waters of the storm into space.

"Countless bodies of water, running or stationary, are constantly giving off moisture from their surface by evaporation. Little brooks hastening to unite and form a rivulet, rivulets contributing their waters to make a river, rivers emptying into larger rivers, and these in turn discharging their floods into the sea; lakes, ponds, marshes, stagnant pools — all, absolutely all, even to the smallest puddle no larger than the hollow of one's hand, yield vapor to the atmosphere without a moment's pause.

"To picture to oneself the total volume of water thus rising continually into the air would be beyond the power of the imagination; and yet that is a mere nothing, for we are forgetting the chief source of atmospheric humidity. We

160

forget the sea, the immense sea, covering three quarters of the earth's surface — the prodigious sea, in comparison with which all the rivers combined are as a mere trifle. What is a drop of water compared with a mill-pond? Nothing. It is the same with the waters irrigating the continents when we contrast their combined volume with the vastness of the sea.

"From the oceans and the bodies of water on the various continents there rises unceasingly into the atmosphere an inconceivable quantity of vapor. Now, this does not stay long over the bodies of water that gave it forth; the wind carries it away, to-day in one direction, to-morrow in another, sometimes to immense distances, so that a layer of air saturated with moisture a thousand leagues from here may reach us and furnish the air we breathe.

"Through the agency of storms, hurricanes, and all kinds of winds, gentle or boisterous, that agitate the air, there is brought about an indiscriminate redistribution, in every direction, of the vapor that rose from any given part of sea or land. In this way the air about us holds moisture, always and everywhere.

"Yes, this air that surrounds us now, this air in which we come and go, contains water in the form of invisible vapor; and it always contains it, sometimes more, sometimes less, every moment and in all seasons. You will be convinced of this if I succeed in making visible to you what is now invisible; if, in short, I succeed in reconverting into water, running water, this vapor which no eye can see.

"As it required only a slight rise in temperature to convert ordinary water into invisible vapor, it will suffice to lower the temperature a little, and thus cool the vapor, to restore it to its former condition, or in other words to reconvert it into water. This cooling process, this taking away of heat, will undo what warming or the application of heat had done in the first place. That is all plain enough, it seems tome.

"Let us go back a moment to the pot of boiling water. Cover it with the lid, first wiping the latter perfectly dry inside; or, better still, hold the lid over the steam escaping from the pot, and at a short distance from the mouth.

"You foresee what will happen. The inside of the lid, perfectly dry at first, will in a few moments be covered with drops of water. Where do these drops come from if not from the steam which, by contact with the cold lid, has lost its heat, to which it owed its subtle form, and has returned to its original state, that of water? So much, then, we have demonstrated, that cold changes vapor to water; which is exactly the opposite of the change wrought by heat.

"For the sake of brevity we will in our future talks use a few terms which I will here define. The return of vapor to the state of water is called condensation. The opposite term is evaporation, which designates the conversion of water into vapor. If we wish to say that vapor becomes water again, we say it condenses.

"How shall we make the invisible vapor that is in the air manifest itself as water? Nothing easier, if we have some ice or snow at our disposal. We fill a water-bottle with small pieces of ice, wipe the outside well to remove any

161

moisture that may already be there, and put the bottle on a perfectly dry plate.

"We do not have to wait long for the result. The glass of the water-bottle, at first perfectly clear, becomes dull and veiled in a kind of fog. Then little drops form, grow bigger, and slowly run down into the plate. Wait a quarter of an hour and we shall have enough water in the plate to pour into a glass and drink if we wish.

"Where does this water come from, if you please? Certainly not from the inside of the water-bottle, for water cannot pass through glass. Then it must come from the surrounding air, which by contact with the ice-cooled glass has itself become so chilled as to make its vaporous contents first appear as a kind of fog, the initial step in condensation, and then run down the glass in the form of drops. In this manner it may be proved at any season of the year that there is a certain amount of moisture in the air.

"We do not always have ice or snow at our disposal, especially in summer. During that season must we, then, forego this interesting and instructive experiment? Certainly not. We merely have to fill the water-bottle with very cold water and we shall see the glass become dull and clouded with moisture.

"There may even be some formation of drops large enough to run down the sides. The result will be the same as when ice is used, only less pronounced, because the cooling effect of cold water is less than that of ice.

"This phenomenon may have been presented to our view many times without attracting our notice. The carafe of cold water placed on our table at dinner very soon loses its transparency and turns dull from the condensation of vapor on the outside. A glass filled with cold water ceases to be transparent, becomes covered with a dull cloud, and looks as if badly washed. That, too, is the vapor from the surrounding air condensing — and collecting on the cold object."

Chapter Sixty-One - Rain

"WHAT is needed to make the half -condensed vapor of clouds finish condensing and turn to water, falling in drops of rain? Very little: a slight cooling, a breath of cold air from some less heated region.

"As soon as it is cooled the fine aqueous dust of the clouds, similar to what we see rolling up in the form of fog, collects in very small drops which fall of their own weight.

"In passing through the dense mist below them these droplets condense on their surface a little of the vapor they meet with, so that they increase in size and become drops capable of growing still larger, until the very moment of their leaving the clouds. Finally they fall to the ground after attaining a size proportioned to the thickness of the cloud-layer through which they have passed.

"That is rain — one day a fine shower which hardly bends the blades of grass, another day a heavy downpour in big drops that patter on the foliage and the tiled roofs. That, I say, is rain which we all so eagerly desire when the country is suffering from drought.

"It rains beneath the clouds that are converted into water by being chilled; but elsewhere it does not rain, the sky is blue, and the sun shines. One can often see this uneven distribution of a rainfall. Has it not happened to you that, with the sky all blue overhead, you have seen in the distance something like a vast grayish curtain, vertically striped and reaching from the sky to the earth? That is a cloud turning into rain and pouring out its contents wherever it passes. It may even have reached you, driven in your direction by the wind. Then the blue sky has suddenly become somber, and you are caught in a shower.

"Clouds may be likened to immense celestial watering-pots that travel almost everywhere at the caprice of the winds that drive them. Every region they visit receives a shower; but any other, however near it may be, receives nothing so long as it does not lie under them. A rainfall may be local in extent and its limits so sharply defined that a few steps in one direction will expose you to a shower and a few steps in the other place you where not a drop falls. But local rains are not the only ones. There are some — and they often occur — that embrace enormous regions, several provinces at a time.

"While the rain is falling, as we may imagine it to fall, let us consider for a moment the marvelous journey accomplished by a single raindrop. Whence does it come? From the clouds floating there above our heads, perhaps one or two thousand meters high. It was up there when the thunder burst; it was present at the blinding flash of the lightning. But no sooner was it formed than, by the force of its own weight, it fell with dizzy speed from its lofty position. Behold it rebounding from a leaf and falling to the ground, where it soaks into the soil and adds to the moisture essential to the life of every plant. A head of lettuce, perhaps, in drinking it, will gain renewed vigor.

"It came from the clouds, and clouds are made of vapor held in the atmosphere. This in turn results from the evaporation of water, chiefly the water of the sea, by the heat of the sun. But of what sea? Who can say? Who could indicate precisely from what point of the broad ocean's surface the sun drew the vapor that was one day to form that drop? Was it from the blue waves of the Mediterranean, the smiling sea to the south of France? It is possible, if the cloud from which it fell was driven so far by the south wind. Or was it from the greenish waves of the ocean whose billows dash furiously against the cliffs of Normandy and the reefs of Brittany? Possibly, if the west wind drove hither the cloud that was to let it fall.

"It is possible, again, that the raindrop came from a far greater distance, perhaps from some gulf fringed with cocoanut-trees whereon perch green parrots with red tails; perhaps from some arm of the sea where the whale suckles its young; perhaps from the other end of the world. Yes, there are all

163

these possibilities; and then what a journey to come to us and water a head of lettuce!

"This prodigious journey ended, in our drop of rain at last to find rest in the plant it has watered? By no means. Nothing remains at rest in this world, not even a drop of water. Everything is in motion, everything is busy, everything is forever beginning again the task already accomplished.

"The drop of water rises with the sap from the roots of the plant, ascends through the stalk, and reaches the leaves, where it evaporates. The heat of the sun reduces to vapor what it had for a moment relinquished to the earth as water. And so we have again the drop of water high in the heavens and transformed to an invisible state — once more given over to the caprices of wind and storm, which will carry it no one knows whither. One day or another it will become rain again, and there is no reason why it should not, sooner or later, water the cocoanut-tree from the neighborhood of which we supposed it to start.

"These journeys being repeated unceasingly, sometimes in one direction, sometimes in another, the raindrop cannot fail some day to rejoin the sea whence it originally came. All rain comes ultimately from the sea, and to the sea all rain finally returns."

Chapter Sixty-Two - Snow

"SNOW has the same origin as rain: it comes from vapor in the atmosphere, especially from vapor rising from the surface of the sea. When a sudden cooling-off takes place in clouds at a high elevation, the condensation of vapor is immediately followed by freezing, which turns water into ice.

"I have already told you [1] that cirrus clouds, which are the highest of all clouds and hence more exposed to cold than the others, are composed of extremely fine needles of ice. Lower clouds, too, if subjected to a sufficient degree of cold, undergo the same transformation. Then there follows a symmetrical grouping of adjacent needles in delicate

Snow Crystals

six-pointed stars which, in greater or less numbers and heaped together at random, make a snowflake. Soon afterward, when it has grown too heavy to float in the air, the flake falls to the ground.

"Examine attentively one that has just fallen on the dark background of your sleeve or cap. You will see a mass of beautiful little starry crystals so graceful in form, so delicate in structure, that the most skilful fingers could never hope to make anything like them. These exquisite formations, which put to shame our poor human artistry, have nevertheless sprung from the haphazard mingling of cloud-masses.

"Such then is the nature of snow, the schoolboy's favorite plaything. From a somber and silent sky it falls softly, almost perpendicularly. The eye follows it in its fall. Above, in the gray depths, it looks like the confused whirling of a swarm of white insects; below it resembles a shower of down, each flake turning round and round and reaching the ground only after considerable hesitation. If the snowfall continues thus for a little while, everything will be hidden under a sheet of dazzling whiteness.

"Now is the time for dusting the back of a schoolmate with a well-directed snowball, which will bring a prompt reply. Now is the time for rolling up an immense snowball which, turning over and over and creaking as it grows, at last becomes too large to move even under our united efforts. On top of this ball a similar one will be hoisted, then another still smaller on that, and the whole will be shaped into a grotesque giant having for mustache two large turkey feathers and for arms an old broomstick. But look out for the hands in modeling this masterpiece! More than one young sculptor will hasten to thrust them, aching with cold, into his pockets. But, though inactive himself, he will none the less give the others plenty of advice on how to finish off the colossus.

"Oh, how glorious is a holiday when there is snow on the ground! If I were to let myself go, how eloquent I could be on the subject! But, after all, what could I say that would be new to you? You know better than I all about the games appropriate to the occasion. You belong to the present, I to the past; you make the snow man now and here; I only tell about it from memory. We shall do better to go on with our modest studies, in which I can be of some help to you.

"From snow to hail is a short step, both being nothing but atmospheric vapor turned to ice by cold. But while snow is in delicate flakes, hail takes the form of hard pellets of ice called hailstones. These vary greatly in size, from that of a tiny pin-head to that of a pea, a plum, a pigeon's egg, and larger.

"Hail often does much harm. The icy pellets, hard as stone, in falling from the clouds gain speed enough to make them break window-panes, bruise the unfortunate person not under cover, and cut to pieces in a few minutes harvests, vineyards, and fruit crops. It is nearly always in warm weather that hail falls, and as necessary conditions there must be a violent storm with flashes of lightning and peals of thunder.

"If on the one hand a hail-storm is to be regarded as a disaster, on the other a fall of snow is often to be welcomed as a blessing. Snow slowly saturates the earth with moisture that is of more lasting benefit than a rainfall. It also covers the fields with a mantle that affords protection from severe frost, so that the young shoots from seeds recently sown remain green and vigorous instead of being exposed to the deadly sting of the north wind.

"Snow plays still another part, and a very important one, a part having to do with the very existence of our streams. On account of the cold in high regions it snows more often on the mountains than in the plains. In our latitude peaks three thousand meters high, or more, are unvisited by rain. Every cloud borne to them by the wind deposits, instead of a shower of rain, a mantle of snow, and that in all seasons of the year, summer as well as winter.

"Driven by the wind or sliding down the steep slopes, this snow from the mountain-tops, renewed almost daily, collects in the neighboring valleys and piles up there in drifts hundreds of meters deep, which finally turn to ice as hard and clear as that of the pond where we go skating. In this way there are formed and maintained those masses of moving ice known as glaciers, immense reservoirs of frozen water which abound in all the larger mountain systems.

"In its upper reaches, where the mountain peaks pierce the sky, the glacier is continually receiving fresh snow that comes sliding down the neighboring slopes, while in its lower course, farther down the valley, where the warmth is sufficient, the ice melts and gives rise to a stream which is soon added to by others from neighboring glaciers. In this way the largest rivers are started on their courses.

"From the soil saturated with rain-water and snow-water come springs and brooks and larger streams, each but a slender thread at first, a few drops trickling slowly, a tiny streamlet that one could stop with the hand. But, collecting drop by drop from all around, trickling down the mountainside, a little here and a little there, one thimbleful added to another, one tiny streamlet uniting with its neighbor, at last we have, first, the little brook babbling over its smooth pebbles, then the larger brook that drives the mill-wheel, then the stream on which rides the rowboat, and finally the majestic river carrying to the sea all the drainage of an immense watershed.

"Every drop of water that irrigates the soil comes from the sea, and every drop returns to the sea. The heat of the sun draws the water up in the form of vapor; this vapor goes to make clouds, which the wind scatters in all directions; from these clouds fall rain and snow; and from this rain and snow are formed rivers and other streams which all combine to return to the sea the water thus distributed.

"The water of every spring, well, fountain, lake, pond, marsh, and ditch — all, absolutely all, even to the tiniest mud-puddle and the moisture that bedews a sprig of moss — comes from the sea and returns to it.

"If water cannot run because it is held back in the hollow of some rock, or in a depression in the ground, or in a leaf that has drunk its fill of sap, no mat-

166

ter: the great journey will be accomplished all the same. The sun will turn it to vapor, which mil be dissipated in the atmosphere; and, having once started on this broad highway that leads everywhere, sooner or later it must return to the sea.

"I hope now you are beginning to understand all this, except one puzzling detail that must certainly have occurred to you. You wonder how it can be that, sea-water being so salt and so disagreeable to drink, rain-water, snow-water, spring-water, river water, and so on, should be so tasteless. The answer is easy. Recall to mind the experiment of the plate of salt water placed in the sun. The part that disappears, evaporated by the heat, is pure water and nothing more. What remains in the plate is the salt that water contained, a substance on which evaporation has no hold.

"The same process of evaporation is constantly going on over all the broad expanse of the sea: the water alone is reduced to vapor, the salt remains. From this vapor, purged of all that made the seawater so disagreeable to the taste, only tasteless water can result."

[1] See "The Story-Book of Science."

Chapter Sixty-Three - Ice

"Do you remember," asked Uncle Paul, "what happened last winter to the garden pump — that handsome red copper pump that shines so in the sun? You have some idea how a pump is made. Below there is a long lead pipe that runs down into the well, while above is a short thick pipe in which the piston goes up and down. This thick pipe is the body of the pump.

"One very frosty morning the pump was found burst from top to bottom; there was a crack as wide as your finger, and out of this crack projected a strip of ice. It caused such a commotion in the house! It was wash-day, and the pump would not work. Mother Ambroisine spent all the morning fetching water from the spring."

"I think," said Jules, "they said the cold had burst the pump; but I racked my brains without being able to understand how cold could burst a metal pipe. An iron or copper pipe is so hard!"

"And then. Uncle," put in Emile, "that same night, the night of the hard frost, I had left my penholder, a small metal tube, on the bench in the garden. I found it again in the morning not damaged in the least. How had it stood the cold when the big strong pipe of the pump had burst?"

"That will all become clear," his uncle replied, "if you will listen to me."

"We are listening with both ears, Uncle Paul," Emile returned, eager for the explanation.

"It is true the cold had burst the pump, but not the cold alone. There was something in the body of the pump, there was –"

"There was water," Jules hastened to interpose.

"This water, when the cold came, was turned into ice which found itself imprisoned between the body of the pump and the piston, without being able to move either up or down. Now, I must tell you that ice expands as it forms. It expands so much that if it happens to be imprisoned it pushes this way and that, in all directions, and breaks the obstacle that arrests its expansion. So the body of the pump burst because ice formed inside."

"Then the tube of my penholder was not burst by the cold because there was no water in it?" asked Emile.

"Exactly."

"But if there had been water there which could not get out?"

"It would most certainly have burst."

"That will be a fine experiment for next winter."

"Speaking of experiments, I will tell you of one that will show you the power of ice when it forms and expands in an enclosed space. What is more solid than a cannon? It is made of bronze, a metal almost as unyielding as iron; it weighs several hundred pounds; and its cylindrical wall is a hand's breadth or more thick. The cannoneer loads it with a small sack of powder and a ball that Emile would find it hard to lift. The powder takes fire, there is an explosion like a clap of thunder, and the iron ball is shot to the distance of a league or even farther. Judge then the resistance this terrible engine of war must offer.

"Well, the expansive force of ice has been tried on cannons. A cannon is filled with water and its mouth stopped up with a solid iron plug screwed so tight that it cannot move. Then the whole is exposed to the cold during a severe winter day. The water freezes and soon the cannon is split from end to end, the ice crowding out through the cracks. What wonder that the pipe of a pump bursts under pressure of ice, when a cannon is rent like an old rag? I must tell you further that this rupture under the pressure of freezing water is accomplished in the quietest way. There is no explosion, as you might imagine there would be, no scattering of flying fragments in every direction. Without any noise the metal is forced apart, and that is all. Should you be sitting astride the cannon, you would have nothing to fear when the rupture came."

The children listened very attentively, being much interested in the bursting of a cannon by something apparently so harmless as ice. But in one particular they were left unsatisfied: they had no opportunity to test for themselves the power of ice. Their uncle read their thoughts, and added:

"There is not much chance of your ever witnessing the bursting of a cannon by ice, and your eyes tell me that you are waiting for me to suggest some substitute. Well, then, how will this do? Next winter take a bottle, fill it full of water, and then cork it securely with a good stopper tied down with string. Put your bottle out of doors when there is a sharp frost. Sooner or later you will find it in pieces, broken by the pressure of the ice. Here again there is no danger. The pieces of the bottle are not sent flying all about, but remain close together, clinging to the ice; or else they fall harmlessly to the ground. That will be an experiment more worthwhile than the one with the penholder. Try it when winter comes."

"We certainly shall," Jules responded. "It will be a curious sight. I have an idea. Uncle Paul; let me tell it to you. In the new pump that was put in to take the place of the old one when that was burst, there is a tap at the bottom; and when it seems likely that there will be a hard frost Mother Ambroisine always goes and opens the tap to let the water run out. That must be to keep ice from forming inside the pump?"

"Yes, that is it. Moreover, as one might forget to open the tap, it is prudent during very cold weather to wrap the pump with rags or straw to shelter it from the air and prevent it getting too cold. That is a precaution to be taken next winter."

Chapter Sixty-Four - Pebbles

WHAT might not a stone teach us if we could only make it tell its story; or, rather, if we could read what it offers to eyes that know how to decipher the inscription! Perhaps some very interesting facts. A pebble has so long a life! It is as old as the world. It has witnessed events of the remotest antiquity, but is a silent witness and guards its secrets so well that we can get little clue to them even by the most attentive study. Nevertheless let us attempt this study.

"We will go down to the river nearby, where the water runs over a bed of stones that are worn smooth as if some patient marble-cutter had taken it into his head to polish them. There are some almost as round as balls, others oval, still others flat, some large and some small, some short and some long, and others that are white, ash-colored, gray, black, or reddish. The smallest look not unlike the sugarplums that fill the confectioner's jars. All are remarkable for their polished surface. Their smooth outlines are pleasing to the touch. These waterworn stones found in our streams are known as pebbles or pebble-stones.

"Who fashioned and polished them in this manner! A fragment of stone such as is broken from a rock by some chance does not look at all like this. It is irregular, angular, rough to the touch. On the rocky slopes of mountains are seen only stones as shapeless as those that the road-maker leaves piled up along the way after he has broken them. Why, then, when under water are they always sleek and round? The answer is not far to seek.

"As a result of the disturbance created by storms, by the melting of great masses of snow, and by violent freshets, the river receives a vast quantity of loose stones that have been swept down from the neighboring slopes. These stones are at first shapeless, with the sharp edges and the many irregularities of stones broken by chance. Henceforth they are under water in the bed of the river. What will be the result? You will know if you stop to think what would take place if a multitude of little irregular pieces of stone were shaken for a long time in a rolling cask.

"Falling against one another unceasingly, clashing together and in constant friction, these pieces would gradually lose their sharp corners, tone down their little roughnesses, and end by becoming perfectly smooth. Marbles, which you prize so highly, are rounded and polished in this way. Small pieces of stone are first roughly shaped with a hammer, then thrown into a rolling cask, and there the work is finished and brought to perfection.

"Running water plays the same part as the rolling cask. At the season of high water the force of the current displaces the stones lying at the bottom of the stream and carries them away, rolling them long distances. Continually dashed one against another, these stones gradually become rounded, after which the loose sand carried away with them rubs and smooths them; and lastly the fine mud that is washed over them again and again imparts, by its gentle action, the final polish. But work of this sort takes a long time. If months do not enable the river to fashion its pebble-stones, it will take years; if years are not enough, it will take centuries; for in this matter time is of no account.

"That is how every stream, from the largest to the smallest, strews its bed with rounded pebbles, often in vast quantities. Running water has rolled them along while shaping them; it has carried them sometimes long distances, so that in order to find stones of like nature you would have to go back to the mountains where the brook or the river had its rise. There only would be found the rock that had yielded the fragments destined to become the pebbles of the plain.

"What, you may ask, is the use of this history of pebble-stones? You will see. Each of us may have chanced to notice, far from any stream, either in a flat country or on some hillside or even on some considerable height, great piles of round, polished pebbles similar in every way to those rolled down by rivers. In each instance there is the same smooth surface, the same rounded shape like that of a ball or an egg or a big sugar-plum or a disk. The stones we select one by one on the river-bank in order to make them skim along the surface of the water are not better shaped.

"What, then, has given to these pebbles the form they wear? Evidently running water, for they do not differ from the pebbles found in brooks and rivers. Water has washed them thither, polishing them on the way by mutual friction — the same process that goes on in every river-bed. There is no doubt about that: their rounded form tells the story of these stones very plainly.

"But to-day these piles of pebbles, covering large tracts of land to a great depth, occupy regions that in many instances have not even the tiniest brooklet. Where formerly rivers ran and impetuous torrents roared, there is now nothing but dry land. The streams have all disappeared and their beds alone remain, sometimes several leagues wide, like those of the largest rivers of the world.

"History makes no mention of these ancient streams. Nor can it speak of them, for it is doubtful whether man ever saw them; and if he saw them, the centuries have effaced all remembrance of them. Trails of pebbles are the only witnesses that tell us where these streams once ran.

"Now these pebble-trails often occupy steep slopes and lofty heights that rise far above the surrounding plains. Never could rivers have run over such heights. A stream must have for its bed some ravine, not a ridge of hills. How, then, can we reconcile these two contradictory facts, that running water has certainly been there, as proved by the multitude of smooth pebbles still remaining, and that water could not have reached such heights even in the greatest floods?

"The contradiction disappears and everything is explained when we consider that the earth's surface is subjected to constant variation. Time works changes in all things, even in mountains. In the course of centuries what was once a valley may become a plain, and what was a plain may rise and form a hill. Earthquakes and the sudden uprising of Monte Nuovo, near Naples, have already furnished us some information on this curious subject; and pebbles furnish us still more. They tell us that the heights they now occupy were formerly plains or valleys where mighty rivers ran; they bear witness that what is now a pebbly mountain-side where one would search in vain for the tiniest spring was in ancient times the bed of a raging torrent; they teach us, in a word, that in ages long past profound upheavals changed the surface of the earth. Such is the strange history that a pebble tells us when we know how to question it."

Chapter Sixty-Five - The Force of Steam

"TO-DAY I will show you a couple of experiments we used to perform for our own amusement when I was of your age. You may like to repeat them yourselves at the first opportunity. They will teach you how it is that steam has become the mightiest toiler man has ever had in his service.

"We have seen how the vapor drawn up by the sun's heat forms clouds, which let fall rain and snow, the source of all the streams that flow on our planet. Let us try to understand by what device man has succeeded in harnessing the steam produced by the heating of water, so that he now commands a servant of unlimited strength and fit for all sorts of tasks.

"Take a small bottle, a vial hardly larger than an egg, and pour into it a little water, let us say about a spoonful; then stop the bottle with a good cork so as to make it air-tight. After you have done this put your bottle before the fire, on the warm ashes and not far from the red-hot coals. Then retire some distance, to be out of range of any possible spattering of hot water, and await results.

171

"The water heats, gives forth steam, and finally boils. All at once there comes a *pop*, and up goes the cork into the air, propelled by the thrust of the imprisoned steam. To make this water-pistol go off with the best effect, the stopper must fit exactly and let no steam escape prematurely. If allowed to escape gradually as it formed, in the manner familiar to you in the ordinary tea-kettle, the steam would accomplish nothing whatever.

"But the cork stopper, if it fits properly, completely closes the mouth of the bottle until the steam has acquired sufficient volume and force to drive it out with a sharp report. It might even offer too great resistance, and in that case the bottle itself would have to give way, shattered by the irresistible pressure of the steam.

"Therefore this experiment should be conducted with some caution; otherwise it might cause bums from scalding steam and wounds from broken glass. Keep at a safe distance, then, while the water is heating, but without allowing yourselves to be unduly alarmed, for the danger is really very slight, in fact almost negligible. Unless the cork be tied down it will be almost sure to yield before the bottle itself bursts, and there will be nothing to fear.

"But there is still another and more convenient method of performing the experiment, and one that will excite no alarm even in the most timid. You are familiar with a certain form of penholder in two parts, one fitting into the other. Into the shorter section the pen is inserted, while the other, which is much longer, serves as handle and is grasped by the fingers. It also plays the part of case or sheath, being cylindrical in form and sheathing the pen when not in use. Well, now, take one of these little metal cylinders and pour into it a few drops of water; then cut a slice of potato or carrot, making it about as thick as your finger, and through it thrust the open end of your penholder. The sharp, circular edge of the orifice will cut a little round plug which will remain in the end of the holder as a stopper. Your water-pistol is now loaded and ready to be discharged.

"Hold the end opposite the stopper in the flame (But first be sure that your penholder is not made of celluloid, which is very inflammable. — *Translator.*) of a candle, taking care to use a pair of nippers or a split stick of wood for the purpose, to avoid burning your fingers. Soon you will hear a noise inside the penholder, indicating that the water is beginning to boil and steam is forming. Then the pistol will go off, sending the stopper to some distance, while a jet of steam escapes from the cylinder.

"Your little piece of artillery is soon reloaded. A few more drops of water, another potato plug, and there you have the gun ready for a second discharge. In this manner you may keep up as long as you please this miniature cannonade in which water takes the place of gunpowder and a harmless bit of potato serves as cannon-ball.

"After performing these experiments your inquiring mind will seek the cause of this power possessed by steam. This vapor of water is seen by us rising from drying linen with such tranquillity, such lack of all appearance of strength, that no one pays it any heed. From a boiling pot, again, we see it

172

ascending as harmlessly as possible. But in the experiments with the bottle and the penholder it shows itself possessed of the explosive force of gunpowder. Whence does it derive this force?

"Steam may be likened to a spring which, if confined in too restricted a space and pressed down upon itself, exerts a repulsive force on obstacles opposing it, but is no sooner set free than it ceases to act. It is thus that we may conceive of the power of steam: free to expand at will, steam has no appreciable energy; but closely confined and receiving constant additions, it becomes more and more compressed and puts forth increasing efforts to escape, until at last every bond is burst, no matter how strong.

"If the bottle were left uncorked and the penholder unstopped, the steam generated in them would escape freely as fast as it formed. It could not accumulate and thus become a sort of compressed spring pushing against everything in its way. But with the cork or the potato plug doing its part the situation is entirely changed. Confined in too narrow a space, the steam accumulates until finally it gains strength enough to hurl to a distance, and with a loud report, the obstacle opposing it.

"Of course it is to be understood that this pressure of the steam is exerted not merely against the stopper of the bottle and the plug of the penholder; it is exerted with equal force against all parts of the interior of the bottle and of the penholder, but only the point of least resistance yields to the pressure. Less resistant than the glass of the bottle and the metal of the penholder, the cork stopper and the potato plug give way and are projected to a distance. If they offered sufficient resistance, the bottle and the penholder would burst.

"Perhaps I have not yet made sufficiently clear to you the power of steam pent up in a narrow space. There is a plaything dear to youngsters of your age, the elder-wood pop-gun, which will help to make the matter plain. You are cleverer than I, no doubt, in making and operating this toy. Never mind; for the benefit of any that may be uninformed I will describe this famous pop-gun and tell how it is operated.

"First you select from the hedge a suitable piece of elder-wood — a piece as large around as the neck of a bottle, very straight and even, and of about a spanks length. The elder is unusually rich in pith, and this is easily removable; all you have to do is to push it out with a slender stick, whereupon you have a substantial and serviceable tube for the barrel of your gun. The next thing to do is to whittle out a ramrod with your jack-knife. This ramrod is simply a slender piece of wood that will fit into the tube and reach from one end to the other. In order to make it easier to operate, one extremity of the ramrod is left larger than the rest and serves as handle. That finishes the outfit, and the gun is ready to load.

"A wad of frayed tow from some old bit of cord that is untwisting and coming to pieces is chewed and rechewed and formed into a plug, which is then stuck into the pop-gun and pushed through to the farther end with the ramrod. There you have one outlet of the tube tightly sealed, so that nothing, not even air, can pass through. A second plug of tow, similar to the first, is next thrust into the free opening. Then, resting the handle of the ramrod against your chest, you force the rod itself into the gun-barrel until, with a loud *pop*, the forward plug is projected like a bullet. But projected by what?

"What was there, to begin with, in the pop-gun, between the two plugs? Nothing, you will doubtless reply. Nothing visible, certainly; but invisibility does not mean nothingness. There was air, which we cannot see, but which is none the less real matter. This air was confined between two substantial plugs of tow, which allowed no passage through at either end of the gun. At first the air occupied the entire length of the gun-barrel, a space quite sufficient and affording ample elbow-room.

"But now the ramrod comes into play. The rear plug is pushed toward the forward one so that the air between the two is compressed within an ever-shortening canal; and the more the space diminishes, the more this air strives to regain its former volume, just as a spring when pressed down tries to return to its original position. At last the moment arrives when the thrust of the imprisoned air ejects the forward plug, the only part of the gun that can give way.

"So it is that steam, crowded by the heat that generates it into a space too small for it, behaves exactly like the air forced by the ramrod into a shorter and shorter section of the pop-gun. In each instance there is the same striving for more room, the same violent pressure against opposing obstacles."

Chapter Sixty-Six - Sound

"GIVE a light blow to a wine-glass. The glass will ring, giving forth a sound, weaker or stronger, lower or higher, according to the quality and size of the glass. The sound lasts but a moment, and then ceases. Strike the glass once more and while it is still ringing touch the rim with your finger. Instantly all is still again; not a sound from the glass. Why does it ring when struck, and why does it stop ringing at the touch of a finger? Before replying let us experiment with other resonant objects.

"A violin-string twangs on being scraped by the bow or plucked with the finger, and while it thus gives forth its note it is seen to vibrate rapidly. So rapid, indeed, is its vibration that it appears to fill the entire space between its extreme positions, with the result that it presents a swollen appearance in the middle, after the manner of a spindle. With the cessation of its vibration it becomes silent. It also falls silent immediately at the touch of a finger.

"A bell rings on being struck by its clapper, and if observed closely the substance of the bell will be seen to tremble in an unmistakable manner. Place your hand on the bronze and you will experience a disagreeable sensation almost amounting to pain, due to the vibration of the metal. Finally, if your hand continues to rest on the bell, the vibration will cease and with it the sound.

"Let us try something still more remarkable. Take a pin by its pointed end and bring the head very close to a ringing wine-glass or bell. You will hear a rapid succession of little taps. Whence do they come? From the glass or the bell-metal striking the pin a series of quick blows as long as the ringing continues. They come from the lively trembling of the sonorous object.

"It is unnecessary to cite other examples; these three will suffice. They show us that, in order to give forth sound, a wine-glass or a bell or a violin-string — in short, any object whatever — must be made to tremble or vibrate with great rapidity. The sound is heard as long as the vibration continues, and ceases when the vibrating object returns to a state of rest. That explains why the wine-glass and the violin-string stop sounding at the touch of a finger, and why the bell will not ring if you rest your hand against it. The finger in the one instance, the hand in the other, stop the resonant trembling, and in so doing arrest the sound. Motion causes the sound; rest brings silence.

"To this rapid motion back and forth is given the name I have already used, vibration. An object from which comes a sound is in vibration; it vibrates. Each of its backward and forward movements, too rapid for the eye to follow, is a vibration; and the quicker these vibrations, the higher the note sounded; the slower they are, the lower the note. In a word, sound is motion, and its place in the scale measures the rapidity of that motion.

"In order to be heard, this sound, this motion, must reach us. The hand detects it in its own way when, resting on the vibrating bell, it experiences a very disagreeable thrill; the finger becomes conscious of it in a peculiar manner when, touching the violin-string, it feels a ticking sensation. But how does the ear contrive to receive the sound when it is at a distance from the resonant object and apparently in no sort of communication with it?

"At this point let me invite you to join me in a sport very familiar to you all. We will take a big stone and drop it into a calm sheet of water. Around the place where the stone struck the water there is instantly formed a circle, then another and another, and so on indefinitely; and all these circles, described about the same center and as regular as if drawn with a pair of compasses, grow larger and larger, in successive rings, until they die out at a long distance from their common center, if the sheet of water is large enough.

"Now, do those circles really chase one another over the water's surface as they appear to? One would certainly think so. You remember how fast they go, one ring after another, each in apparent haste to catch up with its predecessor. But the pursued always keep clear of the pursuers, and the distance between the rings remains the same. And so the fact is they are not really chasing one another; they are not, in fact, moving at all; but they have that deceptive appearance, and it will not be hard for ns to understand why.

"Let us drop a straw or a dry leaf upon the surface of the water. When one of these concentric circles passes like a wave, the straw or leaf is lifted up, after which the wave goes on and leaves it, and it sinks down again, remaining exactly where it was in the first place. Thus it is proved that the water

175

does not move forward at all, for if it did it would carry with it the straw or leaf on its surface.

"What, then, are those waves? Mere palpitations of the water as, without changing its place, it gently rises and falls, thus producing a succession of alternate billows and furrows which appear to go chasing after one another. Watch a field of wheat when the wind blows. Its surface undulates in waves that seem to move forward although the wheat-stalks remain firmly rooted in the ground. Of like sort is the apparent movement of a sheet of smooth water when disturbed by the fall of a stone.

"The circles on the water and the undulations of the wheat-field explain to us the nature of sound. Every object emitting a sound is in rapid vibration, and each of these vibrations causes a shock to the surrounding air and produces a wave which spreads out in all directions, immediately followed by a second, a third, and countless others, all resulting from as many successive vibrations.

"In the air thus shocked by the vibrating body there takes place exactly what we see in the sheet of smooth water disturbed by a falling stone and in the wheat-field ruffled by the wind. Without changing its place the air undergoes an undulatory movement which is transmitted to great distances. In other words, airwaves are formed which propagate themselves in every direction at once through the atmosphere, thus taking the form, not of circles, but rather of hollow spheres, all having a common center.

"These air-waves are not visible to us, because air itself is invisible; but they are none the less real, just as real as waves of water and undulations of a wheat-field. If the eye cannot see them, the ear can hear them, for it is from them that sound comes. Hence they are called sound-waves. The ear hears when sound-waves reach it from any vibrating object."

Chapter Sixty-Seven - Sound (continued)

"IF there were no air around us, "Uncle Paul went on, "the bell, the wine-glass, and the violin-string would vibrate to no purpose; we should hear nothing, there being no sound for us to hear. Silence would reign even where now is deafening uproar; the lightning's flash would be followed by no thunderclap. Do you ask why? Because in the absence of air sound-waves would no longer be possible and the ear would receive nothing to convey the sense of sound. If you wish for a proof that silence follows when air is removed, I will give you one.

"In the middle of a hollow glass globe of considerable size is hung a small bell. By means of a stopcock communication with the interior of the glass receptacle may be suspended or reestablished at will. At first it is full of air, as is naturally the case with any vessel that we call empty. If we shake it so as to disturb the bell, we hear the latter ring, the enclosed air transmitting the sound-waves beyond the confines of the glass chamber; and this is true even though the stop-cock be closed, because the undulatory movement of the inner air is transmitted through the glass and into the outer air.

"Now let us exhaust the air within by means of an air-pump. To describe this pump would take us too far out of our way and would involve details too difficult for your comprehension. Let us pass on, then, and suppose the glass globe emptied of all the air it at first contained, and the stop-cock closed to prevent the entrance of any air from without.

"In that condition the globe may be shaken as much as we choose and, though we see the bell swinging and its clapper beating the metal, there comes to our ears no sound whatever. The bell does actually move and is struck by the tongue — that is indisputable — but its vibrations are mute because they start no sound-waves for lack of air. Now we open the stop-cock, and the outer air rushes in and fills the globe, whereupon the bell is heard as in the beginning. The fact is established that with air there is sound; without air, silence.

"Let us return for a moment to the concentric circles formed on the surface of a sheet of water by the falling of a stone. These circles, these waves, are seen to make their way far out from the center, and their rate of progress is slow enough to admit of being followed by the eye. If we wished, we could, with a little care and by the help of a watch having a second-hand, ascertain the time consumed by one of these waves in going from any given point to any other, and thus the distance covered in one second. Sound-waves are transmitted in a similar manner, and so we say that sound travels. How far does it travel in a second? That would be something worth finding out. Let us look into the matter a little.

"You are standing, we will say, in full view of a belfry some distance off. You see the bell swing and even make out the clapper as it strikes the metal. When it strikes you know that it makes a sound, and yet you hear nothing immediately. The sound reaches you a little later, when the bell is swinging back for another blow from the clapper. How is it that the eye sees before the ear hears?

"The message to the eye is transmitted by light, that to the ear by sound-waves. Now, light travels with inconceivable speed, and for it the greatest terrestrial distances are as nothing. Its rapidity of movement is comparable with that of electricity along a wire. We see, then, the clapper striking the bell at the very instant of the concussion, for light no sooner starts on its journey than it arrives at its destination.

"But with sound-waves it is a very different matter. They travel faster than the circular ripples on water, but still they take an appreciable length of time in reaching us. Hence they are behindhand from the moment the clapper strikes the bell, and they fall more and more behind the farther they have to go.

"You watch from a distance a hunter about to shoot at a bird. All attention, you wait for the discharge with the lively interest felt by those of your age in the fall of the tiniest chaffinch from its perch. The shot is fired, you see both the flash and the smoke, but the report does not reach you until a little later. The explanation is the same as in the case of the bell. Light, moving with unparalleled rapidity, shows us the flash and the smoke as soon as the discharge takes place, whereas the sound-waves, which are relatively slow of movement, bring us only later the report of the huntsman's fowling-piece.

"Let us suppose exactly one second to intervene between our seeing the flash caused by the explosion of the gunpowder and our hearing the report. Then, if our distance from the hunter be carefully measured, it will be found to be three hundred and forty meters. Accordingly, it has taken sound one second to travel that distance. By similar experiments it has been ascertained that all sound, whether loud or subdued, shrill or the reverse, caused in one way or in another, travels through the air at the same invariable rate of speed, three hundred and forty meters per second.

"At this point it may be interesting to make a practical application of what we have just learned. Suppose we wish to know how far away are the lightning and thunder we have just seen and heard. If the flash of lightning and the peal of thunder reach us at the same time, the electric discharge was very near, since the sound, despite its relatively low rate of speed, shows no such tardiness in reaching us as it would have done had it come a long distance. But usually the thunder follows the lightning after an appreciable interval, indicating that the electric discharge is not very near.

"To ascertain just how near or how far away the lightning is, we simply have to know how many seconds elapse between the flash and the first thunderclap. If we have no watch by which to count the seconds — as, if we are school-children, it is more than likely we shall not have — let us simply count 'One, two, three, four,' and so on, without haste, but also without lagging. In that way we shall be able to measure the time accurately enough.

"Now, then, attention! Watch the storm-cloud yonder. There comes a flash of lightning, and we count: one, two, three, and so on up to twelve. Ha! the thunder at last. Twelve seconds passed between flash and report, and so it must have taken that length of time for the sound to reach us. The point where the discharge took place is therefore distant from us twelve times three hundred and forty meters, or about four kilometers. What do you say? Isn't that simple enough? How easy it is with a little intelligence, to solve problems that at first seemed extremely perplexing! To find out how far away the thunder is we only have to count one, two, three, and so on.

"The circular waves on the water have something else very interesting to tell us. Let us go back to them once more. The sheet of water, an enclosed basin let us say, is bounded by a wall that serves to hold the water in. Watch closely what takes place where water and wall come in contact. The ripples produced by the fall of a stone make their way toward the wall in ever-widening circles, and reach the wall one after another. Then, as soon as they touch the wall, they start back again; always preserving their circular form and their fixed distance from one another, they move now in the opposite direction.

"Thus the surface of the water becomes ruffled with two series of waves, one moving toward the wall, the other away from it; and these waves, traveling in contrary directions, meet and pass one another with no confusion or hindrance whatever. To designate this change of direction we say that the waves are reflected by the wall. Those moving toward it are direct waves; those returning from it are reflected waves.

"Confronted by a wall, a rock, or any other obstacle barring their passage, sound-waves behave exactly like water-waves: they are reflected by the obstacle and go back in the opposite direction. Hence it is that we may hear the same sound twice in the same place, with an interval between the first and second hearing. The first is caused by direct waves, the second by reflected waves. When the sound reaches us the second time, we call it an echo.

"In future when you hear your own voice from a distance as if some mischievous sprite were mocking you, you will know that the repetition is produced by some opposing object, some wall or rock or other obstruction, that sends back the sound-waves to you by reflecting them. You first received the soundwaves made by your voice directly, and then you received the soundwaves reflected to you by the obstructing object."

Chapter Sixty-Eight - Light

"LANGUAGE, which is older than science, abounds in expressions whose precise meaning does not always conform with reality. People spoke before observing, and the word sanctioned by usage has sometimes been found untrue in the light of subsequent scientific research. For example, we speak of darting a glance at a person or object, to signify a quick look at that person or object.

"As a matter of fact, when we look at anything do we send forth visual rays from our eyes to procure for us information concerning the thing looked at? Does that which makes us see go out from our eyeballs? Certain terms in common use might imply as much; but to take these terms literally would be a serious mistake.

"In the act of looking nothing goes out from our eyes; on the contrary, all that we see comes to them from the thing seen. Our eyeballs do not send out anything; they receive. We do not dart our glances or looks, but direct them toward the objects we wish to examine; that is, we open our eyes and turn them toward these objects in order to receive that which shows them to us.

"And what do we thus receive? We receive light which, coming from the thing seen, gives us knowledge of it by penetrating to the interior of the eye. The gate by which this light-message enters is the orifice to be seen in the front of the eyeball in the shape of a round black spot. It is called the pupil.

"Light tells us about distant objects somewhat as heat announces the presence of a fire and as sound calls attention to the tinkling bell or the booming cannon. The heat comes from the fire and not from us; the sound comes from the bell or the cannon, not from us. Light, which enables us to see, comes from what is seen and not from us.

"Again, we speak of 'palpable darkness,' meaning profound obscurity. The adjective palpable is applied to anything perceptible to the sense of touch. Strictly speaking, air, although a very tenuous substance, is in reality palpable because one needs only to wave the hand in order to feel its contact, gentle as a light breeze. Mist and fog, which are great masses of vapor, might still more fittingly be called palpable. But can this be said of darkness? Can it be likened to a sort of fog that hides things from us by becoming thick and reveals them again by becoming thin? Is it a material substance, a substance that can be felt, and one that shuts us in like a black veil?

"No, darkness is never palpable; it does not thicken like fog, for it is nothing, absolutely nothing but the absence of light. It has no existence of its own; when darkness descends, nothing really falls from the sky, but rather light has been taken away. Darkness is to light what silence is to sound.

"Sound is a very real thing with its aerial waves that travel from the sonorous body to our ears. Light also is a real thing. With incomparable speed it travels from the luminous object to our eyes. Let the sonorous body cease to vibrate, the luminous object cease to shine, and in the one case we have silence, in the other darkness; in short, two negative quantities.

"Any object, to be visible, must give out light. If it does not, it is by that very fact invisible, however perfect the eye may be. In the same way, a bell that does not ring cannot be heard, no matter how keen the sense of hearing. No light, no vision possible; no sound-waves, no hearing possible.

"It is said, however, that certain animals — the cat in particular — are able to see in complete darkness. If that is your opinion, undeceive yourselves; no animal can see when light is entirely wanting.

"The cat, it is true, has the advantage of us: its large eyes, the pupils of which can contract and almost close when exposed to light of dazzling brilliancy, or enlarge to receive more abundantly the dim light of dark places — its large eyes, I say, enable it to find its way where to duller vision all is utter darkness.

"But in reality there is only partial darkness where the cat finds the little light it needs. If light were entirely wanting the animal would open its big eyes in vain; it would see nothing at all, absolutely nothing; and to find the way about it would have to depend on the long hairs of its mustache, just as the blind man depends on his stick.

"Would you like to see how the cat contrives to regulate the admission of light into its eyes! Watch it in the sun. You will see the pupil reduced to a narrow slit like a black line. In order not to be dazzled by the blinding glare, the animal has nearly closed the passage to the light; it has nearly closed the pupil while leaving the eyelids wide open. Take the cat into the shade. The slit of both eyes enlarges and becomes an oval. Put it in semi-darkness and the oval dilates until it becomes almost a circle; and the dilation increases as the light gets dimmer.

"Thanks to its pupils, which open wide and can thus still receive a little light where to others the darkness is complete, the cat can find its way in the dark and hunt by night still better than by day, as it is then invisible to the mice while it sees them well enough.

"Iron heated white-hot to be hammered on the anvil lights up the black-smith's dark shop. The flame of the lamp lights the interior of our dwellings at night. Any substance, if heated sufficiently, becomes, like the white-hot iron and the lamp-flame, a source of light. The sun is the world's torch, the radiant furnace that gives light, heat, and animation to everything, that has life.

"It is a ball of fire about a million and a half times as large as the earth. Its enormous distance from us — more than thirty millions of leagues — reduces it in our eyes to a disk only a couple of spans across; for the more distant an object is the smaller it appears. Small as its remoteness makes it appear, nevertheless it remains for us the undisputed king of the heavens. Whoever should try to look it in the face would immediately be dazzled and compelled to lower his eyelids.

"The stars, countless in number, are so many suns, comparable with ours in brightness and size; but their distance from us is so prodigious that these colossal stars look like mere points of light. Indeed, the greater number of them are not even visible without the aid of a good telescope. These distant suns light other worlds and take hardly any part in the earth's illumination.

"Our sun is the only dispenser of light to our world. To its rays we owe our day; their withdrawal causes our night.

"Illumined by the sun, objects send back or reflect in all directions the light that reaches them, somewhat as a wall or rock reflects the aerial waves that constitute sound. Reaching our eyes, this reflected light causes us to see what surrounds us; it is a sort of luminous echo comparable with the sonorous echo. The illuminated object thus becomes itself a source of light, but of a borrowed light having its true origin elsewhere.

"We have in the heavens above us a splendid example of this borrowed light. The moon has no light of its own. It is a dark body which becomes bright by reflecting the light from the sun. Its half that faces the sun is bright, the other dark. According to the relative positions of earth, sun, and moon, the last named turns toward us at certain times the whole of its bright half, and then the moon is full; later only a part of this half, which shows us the moon in the shape of a crescent; finally the half that does not get the rays of the sun, and for the time being the moon is invisible although still present in our sky. If it shone by its own light the moon would not have these changing aspects, but would always appear to us as does the sun, in an invariably round and luminous form.

"Every object that arrests light casts a shadow. Hold your hand before the lamp when it is lighted in the evening. On the opposite wall you will see a dark shape, the shadow of your hand. In like manner, when the sun shines, the shadow of a tree, wall, or house can be seen on the ground. What, then, is shadow? It is the space left untouched by light, the latter being intercepted by some obstruction.

"Under the cover of a tree, at the foot of a wall, in the shelter of a rock, the direct rays of the sun are cut off; but the light reflected by neighboring objects, themselves lighted directly, always penetrates there. Hence we have an intermediate state between broad daylight and total darkness, a semi-obscurity which would be total darkness were it not for the light reflected by neighboring objects on which the sun shines. It is partial darkness. Where there is no light, either direct or reflected, we have black night."

The End

Lightning Source UK Ltd.
Milton Keynes UK
UKHW041547240521
384283UK00001B/30